计算机科学与技术丛书

鸿蒙
应用开发教程

李永华◎编著

清华大学出版社
北京

内 容 简 介

HarmonyOS不仅是我国第一款真正意义上的操作系统，也是世界上第一款可以使智能穿戴、车机设备、电视等万物互联互通的操作系统。本书结合HarmonyOS开源应用程序的发展前景、系统特点、功能、开发方法、应用基础进行阐述。基本方法包括Ability框架开发，应用开发入门程序设计；Java UI包括框架概述、组件与布局开发，从功能、方法和实例程序等方面介绍；方舟开发框架基于JS扩展的类Web开发范式和基于TS扩展的声明式开发范式；综合应用案例开发包括系统架构、系统流程、开发环境、开发工具、开发语言、开发实现、测试应用。由于篇幅有限，媒体开发、安全开发、AI开发、网络与连接开发、设备管理开发、数据管理开发、原子化服务等内容在配套资源中提供，以供读者学习。

本书可作为信息与通信工程及相关专业的本科生教材，也可作为从事物联网、创新开发和设计的专业技术人员参考用书。

本书封面贴有清华大学出版社防伪标签，无标签者不得销售。
版权所有，侵权必究。举报：010-62782989，beiqinquan@tup.tsinghua.edu.cn。

图书在版编目(CIP)数据

鸿蒙应用开发教程/李永华编著. —北京：清华大学出版社，2023.1(2024.1重印)
（计算机科学与技术丛书）
ISBN 978-7-302-61920-8

Ⅰ.①鸿… Ⅱ.①李… Ⅲ.①移动终端－应用程序－程序设计－教材 Ⅳ.①TN929.53

中国版本图书馆CIP数据核字(2022)第178345号

责任编辑：崔 彤
封面设计：李召霞
责任校对：申晓焕
责任印制：宋 林

出版发行：清华大学出版社
网　　址：https://www.tup.com.cn，https://www.wqxuetang.com
地　　址：北京清华大学学研大厦A座　　邮　编：100084
社 总 机：010-83470000　　　　　　　　邮　购：010-62786544
投稿与读者服务：010-62776969，c-service@tup.tsinghua.edu.cn
质量反馈：010-62772015，zhiliang@tup.tsinghua.edu.cn
课件下载：https://www.tup.com.cn，010-83470236

印 装 者：三河市人民印务有限公司
经　　销：全国新华书店
开　　本：186mm×240mm　　印　张：23.5　　字　数：527千字
版　　次：2023年1月第1版　　　　　　　　印　次：2024年1月第2次印刷
印　　数：1501～2300
定　　价：79.00元

产品编号：098142-01

前言
FOREWORD

  HarmonyOS 是华为技术有限公司开发的一款全新的、面向万物互联时代的全场景分布式操作系统。HarmonyOS 基于微内核、代码小、效率高、跨平台、多终端、不卡顿、长续航、不易受攻击等特点,在传统的单设备基础上,提出了同一套系统能力、适配多种终端形态的分布式理念,旨在创造一个超级虚拟终端互联的世界,将人、设备、场景有机地联系在一起,能够支持手机、平板、智能穿戴、智慧屏等多种终端设备,提供移动办公、运动健康、社交通信等业务范围,将消费者在全场景生活中接触的多种智能终端实现极速发现、极速连接、硬件互助、资源共享。HarmonyOS 将为我国智能制造产业的发展奠定坚实基础,使未来工业软件的应用更加广泛。

  大学作为传播知识、科研创新的主要机构,为社会培养具有创新思维的现代化人才责无旁贷,而具有时代特色的教材又是培养专业知识的基础,所以教材的重要性不言而喻。本书依据当今信息社会的发展趋势,基于工程教育教学经验,是适合国情、具有自身特色的创新实践教材。

  本书内容由浅入深、先理论后实践,创新思维与实践案例相结合,以满足不同层次的读者需求。同时,本书提供实验代码、视频讲解、教学课件、案例实战、习题答案,供读者自学和提高使用。

  本书的内容和素材主要来源于以下几方面:华为公司官网学习平台;作者近几年承担的教育部和北京市的教育、教学改革项目与成果;作者指导的研究生在物联网方向的研究工作与成果总结;北京邮电大学信息工程专业创新实践,该专业同学基于 CDIO 工程教育方法,实现创新研发,不但学到了知识,提高了能力,而且为本书提供了第一手素材和资料,在此向信息工程专业的同学表示感谢。

  本书的编写得到了华为技术有限公司、江苏润和软件股份有限公司、教育部电子信息类专业教学指导委员会、信息工程专业国家第一类特色专业建设项目、信息工程专业国家第二类特色专业建设项目、教育部 CDIO 工程教育模式研究与实践项目、教育部本科教学工程项目、信息工程专业北京市特色专业项目、北京高等学校教育教学改革项目的大力支持；本书由北京邮电大学教育教学改革项目2022SJJX-A01资助，在此表示感谢！

由于作者水平有限，书中不当之处在所难免，敬请读者不吝指正，以便进一步修改和完善。

李永华于北京邮电大学

2022年4月

目 录
CONTENTS

第 1 章　HarmonyOS 概述 ··· 1

　　▶ 微课视频 45 分钟

　　1.1　HarmonyOS 系统架构 ·· 1
　　1.2　HarmonyOS 系统特性 ·· 3
　　1.3　HarmonyOS 系统安全 ·· 7
　　1.4　HarmonyOS App 结构 ·· 9

第 2 章　应用开发基础 ··· 13

　　▶ 微课视频 128 分钟

　　2.1　开发流程 ··· 13
　　2.2　开发工具 ··· 14
　　　　2.2.1　安装软件工具 ·· 15
　　　　2.2.2　配置开发环境 ·· 16
　　　　2.2.3　运行 Hello World ··· 20
　　2.3　应用开发快速入门 ··· 22
　　　　2.3.1　使用 eTS 语言开发 ··· 23
　　　　2.3.2　使用 Java 语言开发 ·· 25
　　　　2.3.3　使用 JS 语言开发 ··· 29
　　　　2.3.4　可视化开发 ··· 31
　　2.4　DevEco Studio 工程管理 ··· 35
　　　　2.4.1　工程结构 ··· 36
　　　　2.4.2　工程操作 ··· 40
　　　　2.4.3　HarmonyOS 共享包 ·· 46
　　2.5　DevEco Studio 开发方法 ··· 51
　　　　2.5.1　低代码开发 ··· 51
　　　　2.5.2　添加 Ability ··· 63

  2.5.3 添加 JS Component 和 JS Page ···················· 65
  2.5.4 跨设备代码编辑 ···················· 65
  2.5.5 定义 HarmonyOS IDL 接口 ···················· 67
  2.5.6 服务卡片操作 ···················· 72
  2.5.7 使用预览器查看应用效果 ···················· 78
  2.5.8 将 SVG 文件转换为 XML 文件 ···················· 83

第 3 章 Ability 框架开发——基于 Java ···················· 84

▶ 微课视频 255 分钟

 3.1 开发概述 ···················· 84
 3.2 Ability 介绍 ···················· 85
  3.2.1 Page Ability ···················· 85
  3.2.2 Service Ability ···················· 92
  3.2.3 Data Ability ···················· 97
  3.2.4 Intent ···················· 103
  3.2.5 Ability 示例 ···················· 106
 3.3 公共事件与通知开发 ···················· 114
  3.3.1 公共事件开发 ···················· 115
  3.3.2 通知开发 ···················· 121
  3.3.3 IntentAgent 开发 ···················· 126
  3.3.4 后台代理定时提醒开发 ···················· 129
 3.4 后台任务调度和管控 ···················· 134
  3.4.1 短时任务 ···················· 135
  3.4.2 长驻任务 ···················· 136
  3.4.3 托管任务 ···················· 137
 3.5 线程管理开发 ···················· 138
  3.5.1 线程管理开发接口关系 ···················· 139
  3.5.2 线程管理开发步骤 ···················· 140
 3.6 线程间通信 ···················· 146
  3.6.1 概述 ···················· 146
  3.6.2 线程间接口关系 ···················· 147
  3.6.3 线程间通信开发步骤 ···················· 150
 3.7 剪贴板开发 ···················· 153
  3.7.1 剪贴板开发接口关系 ···················· 154
  3.7.2 剪贴板开发步骤 ···················· 157

第 4 章　Java UI 开发 ·· 159

▶ 微课视频 285 分钟

- 4.1 Java UI 框架概述 ·· 159
- 4.2 组件与布局开发 ··· 160
- 4.3 常用组件开发 ·· 166
  - 4.3.1 组件通用 XML 属性 ··· 166
  - 4.3.2 Text ·· 166
  - 4.3.3 Button ··· 172
- 4.4 常用布局开发 ·· 175
  - 4.4.1 DirectionalLayout ··· 176
  - 4.4.2 DependentLayout ·· 181
  - 4.4.3 StackLayout ·· 190
  - 4.4.4 TableLayout ··· 193
  - 4.4.5 PositionLayout ·· 201
  - 4.4.6 AdaptiveBoxLayout ·· 203
- 4.5 自定义组件与布局 ·· 206
  - 4.5.1 自定义组件 ··· 207
  - 4.5.2 自定义布局 ··· 211
- 4.6 动画开发 ··· 218
  - 4.6.1 帧动画 ··· 218
  - 4.6.2 数值动画 ··· 219
- 4.7 可见即可说开发 ··· 229

第 5 章　方舟开发框架（ArkUI）——基于 JS 扩展的类 Web 开发范式 ············ 231

▶ 微课视频 188 分钟

- 5.1 开发概述 ··· 231
- 5.2 JS FA 初步应用 ··· 232
  - 5.2.1 JS FA 概述 ··· 232
  - 5.2.2 JS FA 开发应用 ··· 234
- 5.3 构建用户界面 ·· 238
  - 5.3.1 组件 ·· 238
  - 5.3.2 构建布局 ··· 238
  - 5.3.3 添加交互 ··· 243
  - 5.3.4 动画 ·· 245

5.3.5 事件 …… 248
5.3.6 页面路由 …… 251
5.3.7 焦点逻辑 …… 253
5.4 常见组件开发 …… 254
  5.4.1 Text …… 254
  5.4.2 Input …… 258
  5.4.3 Button …… 263
5.5 动效开发 …… 268
  5.5.1 CSS 动画开发 …… 268
  5.5.2 JS 动画 …… 281
5.6 自定义组件 …… 290
5.7 JS FA 调用 PA …… 292
5.8 使用工具自动生成 JS FA 调用 PA 代码 …… 296

第 6 章 方舟开发框架（ArkUI）——基于 TS 扩展的声明式开发范式 …… 301

微课视频 124 分钟

6.1 基于 TS 扩展的声明式开发范式概述 …… 301
6.2 体验声明式 UI …… 302
  6.2.1 创建声明式 UI 工程 …… 302
  6.2.2 初识 Component …… 304
  6.2.3 创建简单视图 …… 305
6.3 页面布局与连接 …… 316
  6.3.1 构建数据模型 …… 316
  6.3.2 构建列表 List 布局 …… 319
  6.3.3 构建分类 Grid 布局 …… 325
  6.3.4 页面跳转与数据传递 …… 332
6.4 绘图和动画 …… 337
  6.4.1 绘制图形 …… 337
  6.4.2 添加动画效果 …… 345

第 7 章 贪吃蛇小游戏 …… 351

微课视频 10 分钟

7.1 总体设计 …… 351
  7.1.1 系统架构 …… 351
  7.1.2 系统流程 …… 352

7.2 开发环境 ………………………………………………………………… 352
　　7.2.1 开发工具 …………………………………………………………… 352
　　7.2.2 开发语言 …………………………………………………………… 353
7.3 开发实现 ………………………………………………………………… 353
　　7.3.1 UI设计开发 ………………………………………………………… 353
　　7.3.2 程序代码开发 ……………………………………………………… 356
7.4 测试应用 ………………………………………………………………… 358
　　7.4.1 程序调试 …………………………………………………………… 359
　　7.4.2 结果展示 …………………………………………………………… 359
7.5 问题解决 ………………………………………………………………… 359

视频目录
Video Contents

视频名称	时长/分钟	位置
HarmonyOS 概述	13	1.1 节节首
HarmonyOS 系统特性	13	1.2 节节首
HarmonyOS 系统安全	8	1.3 节节首
HarmonyOS App 结构	11	1.4 节节首
开发流程	4	2.1 节节首
开发工具	23	2.2 节节首
应用开发快速入门	37	2.3 节节首
工程管理	33	2.4 节节首
DevEco Studio 开发方法	30	2.5 节节首
Ability 开发概述	2	3.1 节节首
Page Ability	18	3.2.1 节节首
Service Ability	7	3.2.2 节节首
Data Ability	9	3.2.3 节节首
Intent	4	3.2.4 节节首
Ability 示例	13	3.2.5 节节首
公共事件开发	11	3.3.1 节节首
通知开发	7	3.3.2 节节首
IntentAgent 开发	4	3.3.3 节节首
定时提醒开发	7	3.3.4 节节首
后台任务调度和管控	11	3.4 节节首
线程管理开发	12	3.5 节节首
线程间通信	10	3.6 节节首
剪贴板开发	6	3.7 节节首
Java UI 框架概述	3	4.1 节节首
组件与布局开发	9	4.2 节节首
Text 开发	11	4.3.2 节节首
Button 开发	6	4.3.3 节节首
DirectionalLayout 开发	8	4.4.1 节节首
DependentLayout 开发	8	4.4.2 节节首
StackLayout 开发	3	4.4.3 节节首
TableLayout 开发	10	4.4.4 节节首
PositionLayout 开发	3	4.4.5 节节首
AdaptiveBoxLayout 开发	6	4.4.6 节节首
定义组件	15	4.5.1 节节首
定义布局	9	4.5.2 节节首

续表

视频名称	时长/分钟	位置
帧动画开发	5	4.6.1 节节首
数值动画开发	26	4.6.2 节节首
可见即可说开发	4	4.7 节节首
方舟开发概述	5	5.1 节节首
JS FA 初步应用	18	5.2 节节首
构建用户界面	35	5.3 节节首
常见组件开发	34	5.4 节节首
动效开发	54	5.5 节节首
自定义组件	5	5.6 节节首
JS FA 调用 PA	11	5.7 节节首
自动生成 JS FA	21	5.8 节节首
基于 TS 扩展的声明	3	6.1 节节首
体验声明式 UI	31	6.2 节节首
页面布局与连接	52	6.3 节节首
绘图和动画	37	6.4 节节首
贪吃蛇小游戏	10	7.1 节节首

第 1 章 HarmonyOS 概述

HarmonyOS(鸿蒙操作系统)是面向万物互联时代的、全新的分布式操作系统。

1.1 HarmonyOS 系统架构

搭载 HarmonyOS 的设备在系统层面融为一体,形成超级终端,使设备的硬件能力可以弹性扩展,实现设备之间资源共享。对消费者而言,HarmonyOS 能够将生活场景中的各类终端进行整合,实现不同终端设备之间的快速连接、能力互助,匹配合适的设备,提供流畅的全场景体验。

面向开发者实现一次开发,多端部署。HarmonyOS 采用了多种分布式技术,使应用开发与不同终端设备的形态差异无关,从而能够聚焦上层业务逻辑。

一套操作系统可以满足不同能力的设备需求,实现弹性部署。HarmonyOS 采用组件化的设计方案,可根据设备的资源能力和业务特征灵活裁剪,满足不同形态终端设备对操作系统的要求。

HarmonyOS 提供支持多种开发语言的 API,包括 Java、XML(Extensible Markup Language)、C/C++、JS(JavaScript)、CSS(Cascading Style Sheets)和 HML(HarmonyOS Markup Language)。

HarmonyOS 整体遵从分层设计,从下向上依次为内核层、系统服务层、框架层和应用层。系统功能按照"系统→子系统→功能/模块"逐级展开,在多设备部署场景下,HarmonyOS 支持根据实际需求裁剪某些非必要的子系统或功能/模块,如图 1-1 所示。

1. 内核层

内核子系统:HarmonyOS 采用多内核设计,支持针对不同资源受限设备选用适合的操作系统(Operating System,OS)内核。内核抽象层(Kernel Abstract Layer,KAL)通过屏蔽多内核差异,对上层提供基础的内核能力,包括进程/线程管理、内存管理、文件系统、网络管理和外设管理等。

驱动子系统:硬件驱动框架(HDF)是 HarmonyOS 硬件生态开放的基础,提供统一外设访问能力和管理框架。

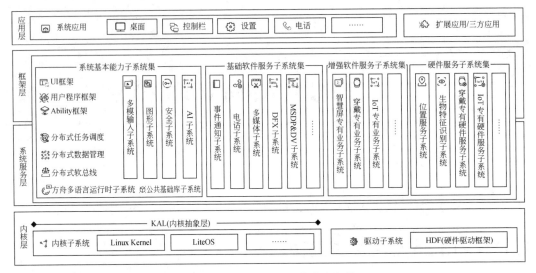

图1-1 HarmonyOS整体技术架构

2. 系统服务层

系统服务层是 HarmonyOS 的核心能力集合,通过框架层对应用程序提供服务。该层包含如下系统集。

系统基本能力子系统集:为分布式应用在 HarmonyOS 多设备上的运行、调度、迁移等操作提供基础能力,由分布式软总线、数据管理、任务调度、方舟运行时、公共基础库、图形、安全、AI 等子系统组成。其中,方舟运行时提供了 C/C++/JS 多语言和基础的系统类库,也为使用方舟编译器静态化的 Java 程序(应用程序或框架层中使用 Java 语言开发的部分)提供运行。

基础软件服务子系统集:公共的、通用的软件服务,由事件通知、电话、多媒体、DFX(Design For X)、MSDP&DV 等子系统组成。

增强软件服务子系统集:针对不同设备的、差异化的能力增强型软件服务,由智慧屏专有业务、穿戴专有业务、IoT 专有业务等子系统组成。

硬件服务子系统集:由位置服务、生物特征识别、穿戴专有硬件服务、IoT 专有硬件服务等子系统组成。

根据不同设备形态的部署环境,基础软件服务子系统集、增强软件服务子系统集、硬件服务子系统集内部可以按子系统粒度裁剪,每个子系统内部又可以按功能粒度裁剪。

3. 框架层

框架层为 HarmonyOS 应用开发提供了 Java/C/C++/JS 等多语言的用户程序框架和 Ability 框架,两种 UI 框架(包括适用于 Java 语言的 Java UI 框架和适用于 JS 语言的 JS UI 框架),以及各种软硬件服务对外开放的多语言框架 API。根据系统的组件化裁剪程度,HarmonyOS 设备支持的 API 也会有所不同。

4．应用层

应用层包括系统应用和第三方非系统应用。HarmonyOS 的应用由一个或多个 FA (Feature Ability) 或 PA(Particle Ability)组成。其中，FA 有 UI(User Interface，用户界面)，提供与用户交互的能力；而 PA 无 UI，提供后台运行任务的能力及统一的数据访问抽象。FA 在进行用户交互时所需的后台数据访问也需要由对应的 PA 提供支撑。基于 FA/PA 的开发应用，能够实现特定的业务功能，支持跨设备调度与分发，为用户提供一致、高效的应用体验。

1.2　HarmonyOS 系统特性

HarmonyOS 系统特性可总结为：硬件互助，资源共享，一次开发，多端部署。多种设备之间能够实现硬件互助，资源共享，依赖的关键技术包括分布式软总线、分布式设备虚拟化、分布式数据管理、分布式任务调度等。HarmonyOS 提供了用户程序框架、Ability 框架及 UI 框架，支持应用开发过程中多终端的业务逻辑和界面逻辑进行复用，提升跨设备应用的开发效率。

1．分布式软总线

分布式软总线是手机、平板、智能穿戴、智慧屏、车机等分布式设备的通信基座，为设备之间的互联互通提供了统一的分布式通信能力，为设备之间的无感发现和零等待传输创造了条件。开发者只需聚焦于业务逻辑的实现，无须关注组网方式与底层协议。分布式软总线如图 1-2 所示。

典型应用场景如智能家居，在烹饪时，手机可以通过碰一碰和烤箱连接，并将自动按照菜谱设置烹调参数，控制烤箱制作菜肴。与此类似，料理机、油烟机、空气净化器、空调、灯、

图 1-2　分布式软总线

窗帘等都可以在手机端显示并通过手机控制。设备之间即连即用，无须烦琐的配置。另一应用场景如多屏联动课堂，教师通过智慧屏授课，与学生开展互动，营造课堂氛围；学生通过平板完成课程学习和随堂问答。统一、全连接的逻辑网络确保传输通道的高带宽、低时延。

2. 分布式设备虚拟化

分布式设备虚拟化平台可以实现不同设备的资源融合、设备管理、数据处理，多种设备共同形成一个超级虚拟终端。针对不同类型的任务，为用户匹配并选择能力合适的执行硬件，使业务连续地在不同设备间流转，充分发挥不同设备的能力优势，例如显示能力、摄像能力、音频能力、交互能力及传感器能力等。分布式设备虚拟化如图1-3所示。

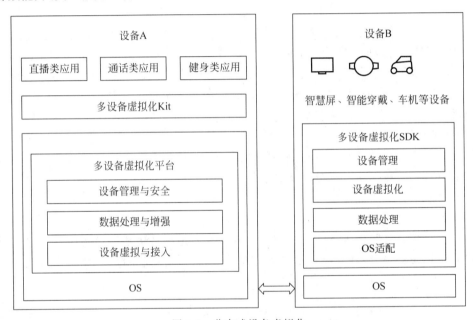

图1-3 分布式设备虚拟化

典型应用场景如视频通话，在做家务时接听视频电话，可以将手机与智慧屏连接，并将智慧屏的屏幕、摄像头与音箱虚拟化为本地资源，替代手机自身的屏幕、摄像头、听筒与扬声器，实现一边做家务、一边通过智慧屏和音箱进行视频通话。另一应用场景如游戏，在智慧屏上玩游戏时，可以将手机虚拟化为遥控器，借助手机的重力传感器、加速度传感器、触控能力，为玩家提供更便捷、更流畅的游戏体验。

3. 分布式数据管理

分布式数据管理基于分布式软总线的能力，实现应用程序数据和用户数据的分布式管理。用户数据不再与单一物理设备绑定，业务逻辑与数据存储分离，跨设备的数据处理如同本地数据处理一样方便快捷，开发者能够轻松实现全场景、多设备下的数据存储、共享和访问，为打造一致、流畅的用户体验创造基础条件。分布式数据管理如图1-4所示。

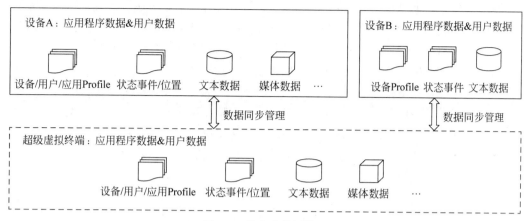

图 1-4　分布式数据管理

典型应用场景如协同办公,将手机上的文档投屏到智慧屏,在智慧屏上对文档执行翻页、缩放、涂鸦等操作,文档的最新状态可以在手机上同步显示。另一应用场景如家庭出游,一家人出游时,妈妈用手机拍的照片,通过家庭照片共享,爸爸可以在自己的手机上浏览、收藏和保存这些照片,家中的爷爷奶奶也可以通过智慧屏浏览。

4. 分布式任务调度

分布式任务调度基于分布式软总线、分布式数据管理、分布式 Profile 等技术特性,构建统一的分布式服务管理(发现、同步、注册、调用)机制,支持对跨设备的应用进行远程启动、远程调用、远程连接及迁移等操作,能够根据不同设备的能力、位置、业务运行状态、资源使用情况,以及用户的习惯和意图,选择合适的设备运行分布式任务。图 1-5 以应用迁移为例,简要地展示了分布式任务调度。

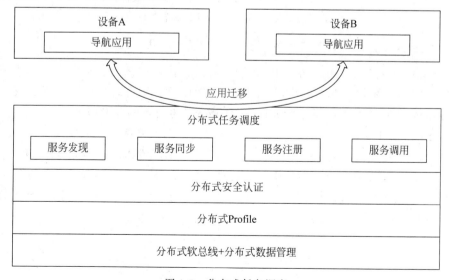

图 1-5　分布式任务调度

典型应用场景如导航,如果用户驾车出行,上车前,在手机上规划好导航路线;上车后,导航自动迁移到车机和车载音箱;下车后,导航自动迁回手机。如果用户骑车出行,在手机上规划好导航路线,骑行时手表可以接续导航。再如外卖场景,在手机上点外卖后,可以将订单信息迁移到手表上,随时查看外卖的配送状态。

5. 一次开发、多端部署

一次开发、多端部署如图 1-6 所示。其中,UI 框架支持 Java 和 JS 两种开发语言,并提供丰富的多态控件,可以在手机、平板、智能穿戴、智慧屏、车机上显示不同的 UI 效果。采用业界主流设计方式,提供多种响应式布局方案,支持栅格化布局,满足不同屏幕的界面适配能力。

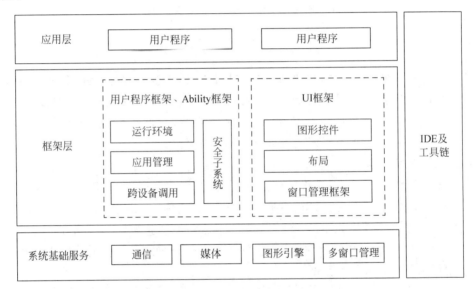

图 1-6 一次开发、多端部署

6. 统一操作系统,弹性部署

HarmonyOS 通过组件化和小型化等设计方法,支持多种终端设备按需弹性部署,能够适配不同类别的硬件资源和功能需求。支撑通过编译链关系自动生成组件化的依赖关系,形成组件树依赖图,支撑产品系统的便捷开发,降低硬件设备的开发门槛。

支持各组件的选择(组件可有可无):根据硬件的形态和需求,可以选择所需的组件。支持组件内功能集的配置(组件可大可小):根据硬件的资源情况和功能需求,可以选择配置组件中的功能集,例如选择配置图形框架组件中的部分控件。支持组件间依赖的关联(平台可大可小):根据编译链关系,可以自动生成组件化的依赖关系,例如选择图形框架组件,将会自动选择依赖的图形引擎组件等。

1.3 HarmonyOS 系统安全

在搭载 HarmonyOS 的分布式终端上，可以保证"正确的人，通过正确的设备，正确地使用数据"。通过"分布式多端协同身份认证"保证"正确的人"；通过"在分布式终端上构筑可信运行环境"保证"正确的设备"；通过"分布式数据在跨终端流动的过程中，对数据进行分类分级管理"保证"正确地使用数据"。

1. 正确的人

在分布式终端场景下，"正确的人"是指通过身份认证的数据访问者和业务操作者。"正确的人"是确保用户数据不被非法访问、用户隐私不泄露的前提条件。HarmonyOS 通过以下三方面实现协同身份认证。

零信任模型：基于零信任模型，实现对用户的认证和对数据的访问控制。当用户需要跨设备访问数据资源或者发起高安全等级的业务操作（安防设备）时，HarmonyOS 会对用户进行身份认证，确保其可靠性。

多因素融合认证：通过用户身份管理，将不同设备上标识同一用户的认证凭据关联，用于标识一个用户，提高认证的准确度。

协同互助认证：通过将硬件和认证能力解耦（信息采集和认证可以在不同的设备上完成），实现不同设备的资源池化及能力的互助与共享，使高安全等级的设备协助低安全等级的设备完成用户身份认证。

2. 正确的设备

在分布式终端场景下，只有保证用户使用的设备是安全可靠的，才能保证用户数据在虚拟终端上得到有效保护，避免用户隐私泄露。

安全启动：确保源头每个虚拟设备运行的系统固件和应用程序是完整的、未经篡改的。通过安全启动，各设备厂商的镜像包就不易被非法替换为恶意程序，从而保护用户的数据和隐私安全。

可信执行环境：提供基于硬件的可信执行环境（Trusted Execution Environment，TEE）保护用户个人敏感数据的存储和处理，确保数据不泄露。由于分布式终端硬件的安全能力不同，对于用户的敏感个人数据，需要使用高安全等级的设备进行存储和处理。HarmonyOS 使用基于数学可证明的形式化开发和验证的 TEE 微内核，获得商用 OS 内核 CC EAL5+ 的认证评级。

设备证书认证：支持为具备可信执行环境的设备预置证书，用于向其他虚拟终端证明自己的安全能力。对于有 TEE 环境的设备，通过预置公钥基础设施（Public Key Infrastructure，PKI）给设备身份提供证明，确保设备是合法制造生产的。在产线进行预置，将设备证书的私钥写入并安全保存在 TEE 环境中，且只在 TEE 内进行使用。在必须传输用户的敏感数据（密钥、加密的生物特征等）时，会在使用设备证书进行安全环境验证后，建立从一台设备的 TEE 到另一台设备的 TEE 之间的安全通道，实现安全传输，如图 1-7 所示。

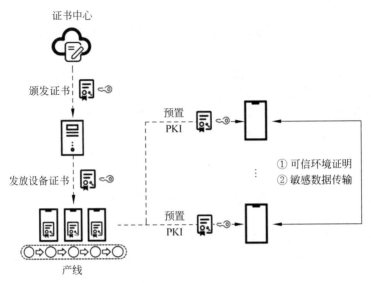

图 1-7 使用设备证书

3. 正确使用数据

在分布式终端场景下,需要确保用户能够正确使用数据。HarmonyOS 围绕数据的生成、存储、使用、传输及销毁过程进行全生命周期的保护,从而保证个人数据与隐私及系统的机密数据(密钥)不泄露。

数据生成:根据数据所在的国家或组织的法律法规与标准规范,对数据进行分类分级,并且根据分类设置相应的保护等级。每个保护等级的数据从生成开始,在其存储、使用、传输的整个生命周期都需要根据对应的安全策略提供不同强度的防护。虚拟超级终端的访问控制系统支持依据标签的访问控制策略,保证数据只能在可以提供足够安全防护的虚拟终端之间存储、使用和传输。

数据存储:通过区分数据的安全等级,存储到不同安全防护能力的分区,对数据进行安全保护,并提供密钥全生命周期的跨设备无缝流动和跨设备密钥访问控制能力,支撑分布式身份认证协同、分布式数据共享等业务。

数据使用:通过硬件为设备提供可信执行环境。用户的个人敏感数据仅在分布式虚拟终端的可信执行环境中进行使用,确保用户数据的安全和隐私不泄露。

数据传输:为了保证数据在虚拟超级终端之间安全流转,需要各设备是正确可信的,建立了信任关系(多个设备通过华为账号建立配对关系),并能够在验证信任关系后,建立安全的连接通道,按照数据流动的规则,安全地传输数据。当设备之间进行通信时,需要基于设备的身份凭据对设备进行身份认证,并在此基础上,建立安全的加密传输通道。

数据销毁:销毁密钥即销毁数据。数据在虚拟终端的存储,都建立在密钥的基础上。当销毁数据时,只需要销毁对应的密钥即完成了数据的销毁。

1.4 HarmonyOS App 结构

用户应用程序泛指运行在设备的操作系统之上,为用户提供特定服务的程序,简称应用。在 HarmonyOS 上运行的应用,有两种形态:传统方式需要安装应用和提供特定功能;免安装的应用(原子化服务)。在 HarmonyOS 中,如无特殊说明,应用所指代的对象包括上述两种形态,类似安卓系统的 App。

1. App 逻辑结构

HarmonyOS 的用户应用程序包以 App Pack(Application Package)形式发布,它由一个或多个 HAP(HarmonyOS Ability Package)及描述每个 HAP 属性的 pack.info 组成。HAP 是 Ability 的部署包,HarmonyOS 应用代码围绕 Ability 组件展开。

一个 HAP 是由代码、资源、第三方库及应用配置文件组成的模块包,可分为 Entry 和 Feature 两种模块类型,如图 1-8 所示。Entry 是应用的主模块,一个 App 中,对于同一设备类型,可以有一个或多个 Entry 类型的 HAP 支持该设备类型中不同规格(API 版本、屏幕规格等)的具体设备。如果同一设备类型存在多个 Entry 模块,则必须配置 distroFilter 分发规则,使应用市场在做云端分发时,对该类型下不同规格的设备进行精确分发。Feature 是应用的动态特性模块,一个 App 可以包含一个或多个 Feature 类型的 HAP,也可以不包含,只有包含 Ability 的 HAP 才能够独立运行。

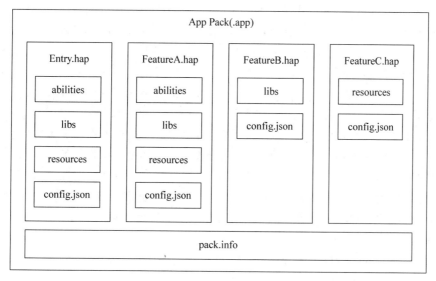

图 1-8 App 逻辑视图

Ability:Ability 是应用所具备的能力抽象,一个应用可以包含一个或多个 Ability,分为 FA 和 PA 两种类型。FA/PA 是应用的基本组成单元,能够实现特定的业务功能。FA 有 UI,而 PA 无 UI。

库文件：库文件是应用依赖的第三方代码（so、jar、bin、har 等二进制文件），存放在 libs 目录下。

资源文件：应用的资源文件（字符串、图片、音频等）存放于 resources 目录下，便于开发者使用和维护。

配置文件：配置文件（config.json）是应用的 Ability 信息，用于声明应用的 Ability 及应用所需权限等信息。

pack.info：描述应用软件包中每个 HAP 的属性，由 IDE 编译生成，应用市场根据该文件进行拆包和 HAP 的分类存储，HAP 的具体属性包括如下内容。

（1）delivery-with-install：表示该 HAP 是否支持随应用安装，True 表示支持随应用安装，False 表示不支持随应用安装。

（2）name：HAP 文件名。

（3）module-type：模块类型，entry 或 feature。

（4）device-type：表示支持该 HAP 运行的设备类型。

HAR（HarmonyOS Ability Resources）：可以提供构建应用所需的所有内容，包括源代码、资源文件和 config.json 文件。HAR 不同于 HAP，它不能独立安装运行在设备上，只能作为应用模块的依赖项被引用。

2. 应用配置文件

应用的每个 HAP 根目录下都存在一个 config.json 配置文件，文件内容主要由 App、deviceConfig 和 module 三部分组成，缺一不可。

App 包括应用的全局配置信息，包含应用的包名、生产厂商、版本号等基本信息。

deviceConfig 包括应用在具体设备的配置信息，包含应用的备份恢复、网络安全等能力。deviceConfig 包含在具体设备上的应用配置信息，可以包含 default、phone、tablet、tv、car、wearable、liteWearable 和 smartVision 等属性。default 标签内的配置适用于所有设备通用，其他设备类型如果有特殊需求，则需要在该设备类型的标签下进行配置。

module 包括 HAP 包的配置信息，包含每个 Ability 必须定义的基本属性（包名、类名、类型和 Ability 提供的能力），以及应用访问系统或其他应用受保护部分所需的权限等。

配置文件 config.json 采用 Json 文件格式，其中包含了一系列配置项，每个配置项由属性和值两部分构成。属性出现顺序不分先后，且每个属性最多只允许出现一次。每个属性的值为 Json 的基本数据类型（数值、字符串、布尔值、数组、对象或者 null 类型）。

3. 资源文件

应用的资源文件（字符串、图片、音频等）统一存放于 resources 目录下，便于开发者使用和维护。resources 目录包括两大类：一类为 base 目录与限定词目录，另一类为 rawfile 目录。

```
resources
|---base    //默认存在的目录
|   |---element
```

```
|   |   |---string.json
|   |---media
|   |   |---icon.png
|---en_GB-vertical-car-mdpi  //限定词目录示例，需要开发者自行创建
|   |---element
|   |   |---string.json
|   |---media
|   |   |---icon.png
|---rawfile  //默认存在的目录
```

base目录与限定词目录按照两级目录形式组织，目录命名必须符合要求，以便根据设备状态匹配相应目录下的资源文件。

一级子目录为base目录与限定词目录。base目录是默认存在的目录。当应用的resources资源目录中没有与设备状态匹配的限定词目录时，会自动引用该目录中的资源文件。base目录与限定词目录下面可以创建资源组目录（element、media、animation、layout、graphic、profile），用于存放特定类型的资源文件。

限定词目录需要开发者自行创建。目录名称由一个或多个表征应用场景或设备特征的限定词组合而成。二级子目录为资源目录，用于存放字符串、颜色、布尔值等基础元素，以及媒体、动画、布局等资源文件。限定词目录可以由一个或多个表征应用场景或设备特征的限定词组合而成，包括移动国家码和移动网络码、语言、文字、国家或地区、横竖屏、设备类型、颜色模式和屏幕密度等维度，限定词之间通过下画线（_）或者中画线（-）连接。开发者在创建限定词目录时，需要掌握命名要求、限定词目录与设备状态的匹配规则。

限定词目录为设备状态匹配对应的资源规则，优先级从高到低依次为：移动国家码和移动网络码→区域（可选组合：语言、语言_文字、语言_国家或地区、语言_文字_国家或地区）→横竖屏→设备类型→颜色模式→屏幕密度。如果限定词目录中包含移动国家码和移动网络码、语言、文字、横竖屏、设备类型、颜色模式限定词，则对应限定词的取值必须与当前的设备状态完全一致，该目录才能够参与设备的资源匹配，例如限定词目录zh_CN-car-ldpi不能参与en_US设备的资源匹配。

rawfile目录支持创建多层子目录，目录名称可以自定义，文件夹内可以自由放置各类资源文件。rawfile目录的文件不会根据设备状态匹配不同的资源，如表1-1所示。

表1-1 资源文件

资源目录	资源文件说明
base > element	包括字符串、整型数、颜色、样式等资源的json文件。每个资源均由json格式进行定义，例如boolean.json为布尔型，color.json为颜色，float.json为浮点型，intarray.json为整型数组，integer.json为整型，pattern.json为样式，plural.json为复数形式，strarray.json为字符串数组，string.json为字符串值

续表

资源目录	资源文件说明
base > graphic	XML 类型的可绘制资源,例如 SVG(Scalable Vector Graphics)为可缩放矢量图形文件、Shape 为基本的几何图形(矩形、圆形、线等)
base > layout	XML 格式的界面布局文件
base > media	多媒体文件,例如图形(.png、)、视频(.gif)、音频(.mp3、.mp4)等
base > profile	用于存储任意格式的原始资源文件。区别在于,rawfile 不会根据设备的状态去匹配不同的资源,需要指定文件路径和文件名进行引用
rawfile	

第 2 章 应用开发基础

本章对 HarmonyOS 使用 DevEco Studio 运行环境、应用程序的开发流程、开发工具和开发基础进行介绍。

2.1 开发流程

针对 HarmonyOS 应用开发流程，使用 DevEco Studio，按照如下步骤，开发并上架 HarmonyOS 应用到华为应用市场，如图 2-1 所示。

1. 开发准备

在进行 HarmonyOS 应用开发前，需要注册一个华为开发者账号，并完成实名认证，认证方式分为"个人实名认证"和"企业实名认证"。

下载 HUAWEI DevEco Studio，一键完成开发工具的安装。开发工具完成后，设置开发环境，对于绝大多数开发者来说，只需要下载 HarmonyOS SDK 即可；如有开发者在企业内部访问 Internet 受限，需要通过代理进行访问，并设置对应的代理服务器才能下载 HarmonyOS SDK。

2. 开发应用

DevEco Studio 集成了 Phone、Tablet、TV、Wearable、LiteWearable 等设备的典型场景模板，可以通过工程向导创建一个新工程，然后定义应用的 UI、开发业务功能等编码工作。在开发过程中，可以使用预览器查看 UI 布局效果，支持实时预览、

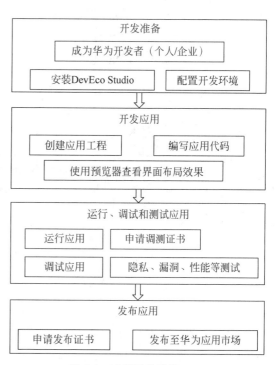

图 2-1 应用开发流程

动态预览、双向预览等功能，使编码的过程更加高效。

3. 运行、调试和测试应用

HarmonyOS 应用开发完成后，可以使用真机或者模拟器进行调试，支持单步调试、跨设备调试、跨语言调试、变量可视化等手段。在发布到应用市场前，还需要对漏洞、隐私、兼容性、稳定性、性能等进行测试，确保 HarmonyOS 应用纯净、安全，给用户带来更好的体验。

4. 发布应用

HarmonyOS 应用开发就绪后，需要布置华为应用市场，以便对其进行分发，普通消费者可以通过应用市场获取到对应的 HarmonyOS 应用。发布到华为应用市场必须使用发布证书进行签名。

2.2 开发工具

HUAWEI DevEco Studio 是基于 IntelliJ IDEA Community 开源版本打造，面向华为终端全场景多设备的一站式集成开发环境(IDE)，为开发者提供工程模板创建、开发、编译、调试、发布等 E2E 的 HarmonyOS 应用开发服务。使用 DevEco Studio，可以更高效地开发具备 HarmonyOS 分布式能力的应用，进而提升创新效率。

作为一款开发工具，除了具有基本的代码开发、编译构建及调测等功能外，DevEco Studio 还具有以下特点，如图 2-2 所示。

图 2-2　DevEco Studio 特点

多设备统一开发环境：支持多种 HarmonyOS 设备的应用开发，包括手机(Phone)、平板(Tablet)、车机(Car)、智慧屏(TV)、智能穿戴(Wearable)、轻量级智能穿戴(LiteWearable)和智慧视觉(Smart Vision)设备。

支持多语言的代码开发和调试：包括 Java、XML(Extensible Markup Language)、C/C++、JS(JavaScript)、CSS(Cascading Style Sheets)和 HML(HarmonyOS Markup Language)。

支持 FA 和 PA 快速开发：通过工程向导快速创建 FA/PA 工程模板，一键式打包成 HAP。

支持分布式多端应用开发：一个工程和一份代码可跨设备运行，支持不同设备界面的实时预览和差异化开发，实现代码的最大化作用。

支持多设备模拟器：提供多设备的模拟器资源，包括手机、平板、车机、智慧屏、智能穿戴设备的模拟器，方便开发者高效调试。

支持应用开发 UI 实时预览：提供 JS 和 Java 预览器功能，可以实时查看应用的布局效果，支持实时预览和动态预览；同时还支持多设备同时预览，查看同一个布局文件在不同设备上的呈现效果。

DevEco Studio 支持 Windows 系统和 macOS 系统，在开发 HarmonyOS 应用前，需要准备 HarmonyOS 应用的开发环境，包括软件安装、配置开发环境和运行 Hello World 共 3 个环节，如图 2-3 所示。

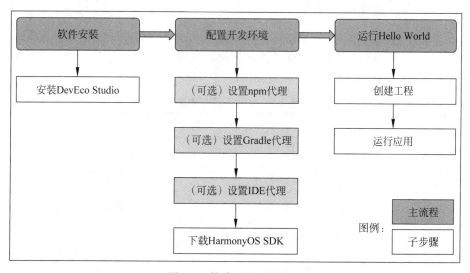

图 2-3 构建开发环境流程

2.2.1 安装软件工具

DevEco Studio 支持 Windows 和 macOS 系统，下面针对这两种操作系统的软件安装方式进行介绍。下载和安装 DevEco Studio，它的编译构建依赖 JDK，DevEco Studio 预置了 Open JDK，版本为 1.8。

1. Windows 环境

为保证 DevEco Studio 正常运行，建议计算机配置满足如下要求：Windows 10 操作系统、64位、8GB及以上内存、100GB及以上硬盘、1280像素×800像素及以上分辨率。

（1）进入HUAWEI DevEco Studio产品页，单击下载列表后的按钮，下载DevEco Studio。下载地址为https://developer.harmonyos.com/cn/develop/deveco-studio#download。

（2）下载完成后，双击下载deveco-studio-xxxx.exe，进入DevEco Studio安装向导，在安装选项界面勾选64-bit launcher后，单击Next按钮，直至安装完成。

2．macOS环境

为保证DevEco Studio正常运行，建议计算机配置满足如下要求。

macOS 10.14/10.15/11.2.2操作系统、8GB及以上内存、100GB及以上硬盘、1280像素×800像素及以上分辨率。

（1）进入HUAWEI DevEco Studio产品页，单击下载列表后的按钮，下载DevEco Studio。下载地址为https://developer.harmonyos.com/cn/develop/deveco-studio#download。

（2）下载完成后，双击下载的deveco-studio-xxxx.dmg软件包。

（3）在安装界面中，将DevEco-Studio.app拖曳到Applications中，等待安装完成。

2.2.2 配置开发环境

DevEco Studio软件安装完成后，检查和配置开发环境。DevEco Studio提供SDK Manager统一管理SDK及工具链，下载各种编程语言的SDK包时，会自动下载SDK包依赖的工具链。SDK Manager提供多种编程语言的SDK包和工具链，如表2-1所示。

表2-1 SDK下载包列表

组件包名	说明
Native	C/C++语言SDK包
eTS	eTS SDK包
JS	JS语言SDK包
Java	Java语言SDK包
System-image-phone	本地模拟器Phone设备镜像文件，仅支持API Version 6
System-image-tv	本地模拟器TV设备镜像文件，仅支持API Version 6
System-image-wearable	本地模拟器Wearable设备镜像文件，仅支持API Version 6
EmulatorX86	本地模拟器工具包
Toolchains	SDK工具链，HarmonyOS应用/服务开发必备工具集，包括编译、打包、签名、数据库管理等工具的集合
Previewer	HarmonyOS应用/服务预览器，在开发过程中可以动态预览Phone、TV、Wearable、LiteWearable等设备的应用/服务效果，支持JS、eTS和Java应用/服务预览

第一次使用DevEco Studio，需要下载HarmonyOS SDK及对应工具链。如果是非首次安装DevEco Studio，则更新HarmonyOS SDK。

1．下载HarmonyOS SDK

下载HarmonyOS SDK步骤如下。

（1）运行已安装的 DevEco Studio，首次使用，请选择 Do not import settings，单击 OK 按钮。

（2）进入配置向导页面，设置 npm registry，DevEco Studio 已预置对应的仓，单击 Start using DevEco Studio 按钮进入下一步，如图 2-4 所示。

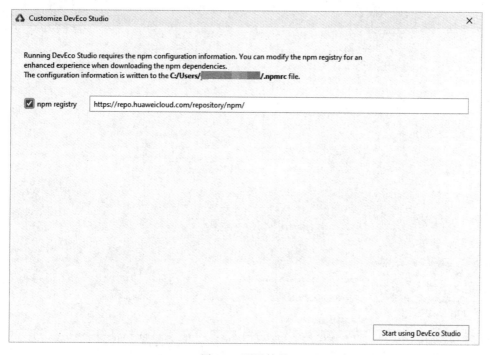

图 2-4　预置结果

（3）设置 Node.js 信息，可以指定本地已安装的 Node.js（版本要求为 v14.19.1 及以上，且低于 v15.0.0）；如果本地没有合适的版本，可以单击 Download 按钮，然后在线下载 Node.js。本示例以下载 Node.js 为例，选择下载源和存储路径后，单击 Next 按钮进入下一步，等待安装完成即可。

（4）在 SDK Components Setup 界面中，设置 OpenHarmony SDK 和 HarmonyOS SDK 存储路径，OpenHarmony SDK 和 HarmonyOS SDK 不能设置为同一个路径，且路径中不能包含中文字符，然后单击 Next 按钮。

（5）在弹出的 SDK 下载信息页面，单击 Next 按钮，并在弹出的 License Agreement 窗口中，单击 Accept 按钮开始下载 SDK。

（6）等待 SDK 及工具下载完成，单击 Finish 按钮，界面会进入 DevEco Studio 欢迎页，如图 2-5 所示。默认下载 API Version 8 的 SDK 及工具链，如需下载其他版本 SDK，请在 HarmonyOS SDK 界面手动下载。

（7）下载完成后，单击 Finish 按钮，HarmonyOS SDK 安装完成，如图 2-6 所示。

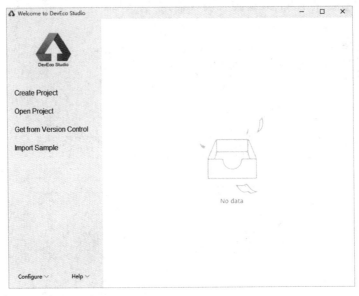

图 2-5　欢迎页

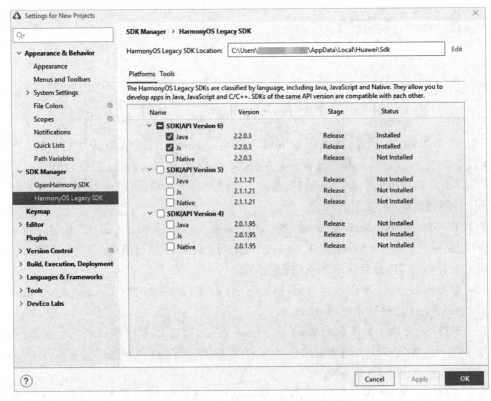

图 2-6　HarmonyOS SDK 安装结果

2. 更新 HarmonyOS SDK

如果已经下载 HarmonyOS SDK,当存在新版本的 SDK 时,可以通过 SDK Manager 更新对应的 SDK,进入 SDK Manager 的步骤如下。

在 DevEco Studio 欢迎页,单击 Configure(或图标)→Settings→SDK Manager→HarmonyOS SDK(macOS 系统为 Configure→Preferences→SDK Manager→HarmonyOS SDK)。

在 DevEco Studio 打开工程的情况下,单击 Files→Settings→SDK Manager→HarmonyOS SDK 进入(macOS 系统为 DevEco Studio→Preferences→SDK Manager→HarmonyOS SDK)。

在 SDK Manager 中,勾选需要更新的 SDK,单击 Apply 按钮,在弹出的确认更新窗口中,单击 OK 按钮开始更新。

3. 配置 HDC 工具环境变量

HDC 提供 HarmonyOS 应用的调试工具,为方便使用 HDC 工具,在端口号设置环境变量,步骤如下。

在此计算机→属性→高级系统设置→高级→环境变量中,添加 HDC 端口变量。变量名为 HDC_SERVER_PORT,变量值设置为 7035,如图 2-7 所示。环境变量配置完成后,关闭并重启 DevEco Studio。

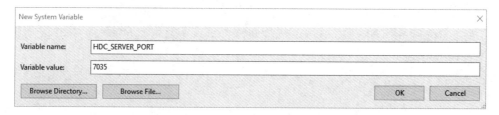

图 2-7　环境变量配置

macOS 环境变量设置步骤如下。

(1) 打开终端工具,执行 vi ./.bash_profile 命令,打开 .bash_profile 文件。

(2) 单击字母 i,进入 Insert 模式。

(3) 输入以下内容,添加 HDC_SERVER_PORT 环境变量信息。

HDC_SERVER_PORT=7035

launchctl setenv HDC_SERVER_PORT $HDC_SERVER_PORT

export HDC_SERVER_PORT

(4) 编辑完成后,按 Esc 键,退出编辑模式,然后输入:wq,按 Enter 键保存。

(5) 执行 source .bash_profile 命令,使配置的环境变量生效。

(6) 环境变量配置完成后,关闭并重启 DevEco Studio。

DevEco Studio 开发环境依赖于网络环境,需要连接网络才能确保工具的正常使用。一般来说,如果使用的是个人或家庭网络,不需要设置代理信息;只有部分企业网络受限的情况下,才需要设置。

2.2.3 运行 Hello World

DevEco Studio 开发环境配置完成后,运行 Hello World 工程验证环境设置是否正确。下面以 Phone 工程为例,在 Phone 的远程模拟器中运行该工程。

1. 创建新工程

创建新工程步骤如下。

(1) 打开 DevEco Studio,在欢迎页中单击 Create Project,创建一个新工程。

(2) 根据工程创建向导,选择需要的 Ability 工程模板,单击 Next 按钮,如图 2-8 所示。

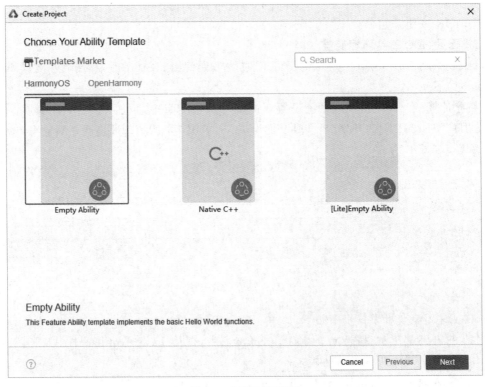

图 2-8 创建工程模板

(3) 填写工程相关信息,主要选择 Compile SDK、Device Type 和 Language,其他保持默认值即可,单击 Finish 按钮,如图 2-9 所示。

(4) 工程创建完成后,DevEco Studio 会自动进行同步,如图 2-10 所示。

2. 使用模拟器运行 Hello World

DevEco Studio 提供模拟器运行和调试 HarmonyOS 应用。

(1) 在 DevEco Studio 菜单栏,单击 Tools→Device Manager。

(2) 在 Remote Emulator 页签中单击 Login,在浏览器中弹出华为开发者联盟账号登录

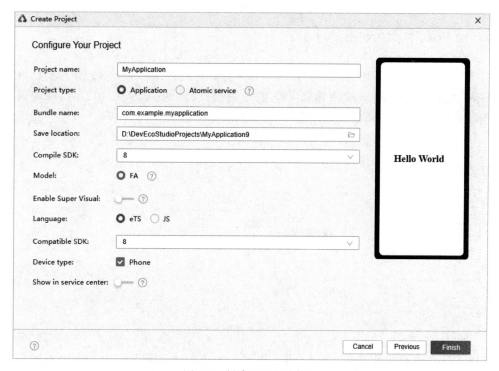

图 2-9　创建工程相关信息

图 2-10　同步成功

界面，输入已实名认证的用户名和密码进行登录。

（3）登录后，单击界面的"允许"按钮进行授权。

（4）在设备列表中，根据 SDK 的版本，选择设备，并单击"运行"按钮，运行模拟器。

（5）单击 DevEco Studio 工具栏中的"运行"按钮或使用默认快捷键 Shift＋F10（Mac 为 control＋R）运行工程。

（6）DevEco Studio 会启动应用的编译构建，完成后应用即可运行在模拟器上。

下面通过一个简单的 DEMO 工程示例，详细介绍开发过程，实现 HarmonyOS 应用开发快速入门。

2.3 应用开发快速入门

本节适合 HarmonyOS 应用开发的初学者学习。通过构建一个简单的具有页面跳转功能的应用熟悉 HarmonyOS 应用开发流程,在左侧的页面中单击 Next 按钮后,跳转到右侧的页面,如图 2-11 所示。

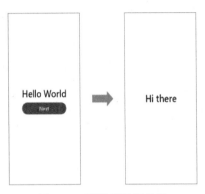

图 2-11　预览器运行效果

HarmonyOS 提供两种 FA 的 UI 开发框架:Java UI 框架和 JS UI 框架。前者提供细粒度的 UI 编程接口,UI 元素更丰富,使应用开发更加灵活;后者提供相对高层的 UI 描述,使应用开发更加简单,Java UI 框架与 JS UI 框架对比如表 2-2 所示。

表 2-2　Java UI 框架与 JS UI 框架对比

比 较 项	Java UI 框架	JS UI 框架
语言生态	Java	JS
接口方式	命令式	声明式
执行方式	开发者处理,基于 API 驱动的 UI 变更	框架层处理,基于数据驱动的 UI 自动变更
系统支持性	只有大型系统支持	覆盖平台更广,轻量系统、小型系统、标准系统、大型系统都支持
相对优势	UI 元素更丰富,开发更灵活	轻量化,开发更简便

另外,DevEco Studio V2.2 Beta1 及更高版本还支持 HarmonyOS 低代码开发方式。低代码开发方式遵循 HarmonyOS JS 开发规范,具有丰富的 UI 编辑功能,通过可视化界面开发方式快速构建布局,可有效降低用户的上手成本,提升用户构建 UI 的效率。

设备类型以 Phone 为例,使用 eTS 语言开发,模板选择 Empty Ability,Language 选择 eTS;使用 Java 语言开发,模板选择 Empty Ability,Language 选择 Java;使用 JS 语言开发,模板选择 Empty Ability,Language 选择 JS;使用可视化开发,模板选择 Empty Ability,Language 选择 JS。工程创建完成后,使用预览器或 Phone 模拟器运行该工程。下面分别用 eTS 语言、Java 语言、JS 语言及可视化方式，实现上述两个页面跳转功能。

2.3.1 使用 eTS 语言开发

eTS 语言开发，使用 DevEco Studio V3.0.0.601 Beta1 及更高版本。模拟器运行时选择 API 7 及以上的设备。若首次打开 DevEco Studio，单击 Create Project 创建工程。如果已有一个工程，单击 File→New→Create Project。选择 HarmonyOS 模板库，选择模板 Empty Ability，单击 Next 按钮进行配置。进入配置工程界面，Language 选择 eTS，其他参数保持默认设置即可。

1. 构建第一个页面

工程同步完成后，在 Project 窗口中，单击 entry→src→main→ets→MainAbility→pages，打开 index.ets 文件，在默认页面基础上，添加一个 Button 组件，作为按钮响应用户单击，从而实现跳转到另一个页面，跳转按钮绑定 onClick 事件，单击按钮时跳转到第二页。index.ets 文件相关代码如下。

```
//index.ets
import router from '@ohos.router';
@Entry
@Component
struct Index {
  @State message: string = 'Hello World'
  build() {
    Row() {
      Column() {
        Text(this.message)
          .fontSize(50)
          .fontWeight(FontWeight.Bold)
        //添加按钮，以响应用户单击
        Button() {
          Text('Next')
            .fontSize(30)
            .fontWeight(FontWeight.Bold)
        }
        .type(ButtonType.Capsule)
        .margin({
          top: 20
        })
        .backgroundColor('#0D9FFB')
        .width('40%')
        .height('5%')
        //跳转按钮绑定 onClick 事件，单击时跳转到第二页
        .onClick(() => {
          router.push({ url: 'pages/second' })
        })
```

```
        }
        .width('100%')
      }
      .height('100%')
  }
}
```

2. 构建第二个页面

新建第二个页面文件。在 Project 窗口中，打开 entry→src→main→ets→MainAbility，右击 pages 文件夹，选择 New→eTS File，命名为 second，单击 Finish 按钮。添加 Text 组件并设置其样式。second.ets 文件示例如下。

```
//second.ets
@Entry
@Component
struct Second {
  @State message: string = 'Hi there'
  build() {
    Row() {
      Column() {
        Text(this.message)
          .fontSize(50)
          .fontWeight(FontWeight.Bold)
      }
      .width('100%')
    }
    .height('100%')
  }
}
```

配置第二个页面的路由，在 config.json 文件中的 module→JS→pages 下配置第二个页面的路由 pages/second。示例代码如下。

```
{
  ...
  "module": {
    "js": [
      {
        ...
        "pages": [
          "pages/index",
          "pages/second"
        ],
        ...
      }
    ]
  }
}
```

            }
        }

3. 实现页面间的跳转

页面间的导航可以通过页面路由 router 实现。页面路由 router 根据页面 URL 找到目标页面，从而实现跳转，即 index.ets 文件的第一行加入如下代码。

```
import router from '@ohos.router';
```

使用预览器或模拟器运行项目，获得跳转效果。

2.3.2 使用 Java 语言开发

Java UI 框架提供了两种编写布局的方式：XML 中声明 UI 布局和在代码中创建布局，二者没有本质差别。下面以 XML 的方式编写第一个页面，通过代码的方式编写第二个页面。

1. XML 方式编写第一个页面

XML 方式编写第一个页面步骤如下。

（1）在 Project 窗口中，单击 entry→src→main→resources→base→layout，打开 ability_main.xml 文件。

（2）（可选）在 Project 窗口中，单击 entry→src→main→resources→base→element，打开 string.json 文件，可在此文件内声明所需引用的资源内容。关于取值为 string 类型的属性，既可以直接设置文本字串，又可以引用 string 资源（推荐使用），string.json 的示例代码如下。

```
{
    "string":[
        {
            "name":"entry_MainAbility",
            "value":"entry_MainAbility"
        },
        {
            "name":"mainability_description",
            "value":"Java_Empty Ability"
        },
        {
            "name":"mainability_HelloWorld",
            "value":"Hello World"
        },
        {
            "name":"button_Next",
            "value":"Next"
        }
    ]
}
```

(3) 第一个页面内有一个文本和一个按钮,使用 DependentLayout 布局,通过 Text 和 Button 组件实现,其中 vp 和 fp 分别表示虚拟像素和字体像素。本示例展示两个组件的显示文本分别采用直接设置文本字符串、引用 string 资源(推荐使用)的方式,ability_main.xml 的示例代码如下。

```xml
<?xml version="1.0" encoding="utf-8"?>
<DependentLayout
    xmlns:ohos="http://schemas.huawei.com/res/ohos"
    ohos:width="match_parent"
    ohos:height="match_parent">
    <Text
        ohos:id="$+id:text"
        ohos:width="match_content"
        ohos:height="match_content"
        ohos:text="Hello World"
        ohos:text_color="#000000"
        ohos:text_size="32fp"
        ohos:center_in_parent="true"/>
    <!-- 无上一步的可选操作,则 Button 组件设置为 ohos:text="Next"  -->
    <Button
        ohos:id="$+id:button"
        ohos:width="match_content"
        ohos:height="match_content"
        ohos:text="$string:button_Next"
        ohos:text_size="19fp"
        ohos:text_color="#FFFFFF"
        ohos:top_padding="8vp"
        ohos:bottom_padding="8vp"
        ohos:right_padding="70vp"
        ohos:left_padding="70vp"
        ohos:center_in_parent="true"
        ohos:below="$id:text"
        ohos:margin="10vp"/>
</DependentLayout>
```

(4) 按钮的背景是蓝色胶囊样式,可以通过 graphic 目录下的 XML 文件设置。右击 graphic 文件夹,选择 New→File,命名为 background_button.xml,按 Enter 键,相关代码如下。

```xml
<?xml version="1.0" encoding="utf-8"?>
<shape
    xmlns:ohos="http://schemas.huawei.com/res/ohos"
    ohos:shape="rectangle">
    <corners
        ohos:radius="100"/>
    <solid
```

```xml
        ohos:color = "#007DFF"/>
</shape>
```

在 layout 目录下的 ability_main.xml 文件中<Button 下加入一行代码，使用 background_element="$graphic:background_button"的方式引用 background_button.xml 文件，其中<Button 部分的代码如下。

```xml
<?xml version = "1.0" encoding = "utf-8"?>
<DependentLayout
    ...
    <Button
        ohos:id = "$+id:button"
        ohos:width = "match_content"
        ohos:height = "match_content"
        ohos:text = "Next"
        ohos:text_size = "19fp"
        ohos:text_color = "#FFFFFF"
        ohos:top_padding = "8vp"
        ohos:bottom_padding = "8vp"
        ohos:right_padding = "70vp"
        ohos:left_padding = "70vp"
        ohos:center_in_parent = "true"
        ohos:below = "$id:text"
        ohos:margin = "10vp"
        ohos:background_element = "$graphic:background_button"/>
</DependentLayout>
```

(5) XML 文件中添加组件后，需要在 Java 代码中加载 XML 布局。在 Project 窗口中，选择 entry→src→main→java→com.example.myapplication→slice,选择 MainAbilitySlice.java 文件，使用 setUIContent 方法加载 ability_main.xml 布局。此外，运行代码前需采用 import(可使用 Alt+Enter 快捷键)引入对应类，否则会报错提示，MainAbilitySlice.java 的示例代码如下。

```java
//根据实际工程/包名引入
package com.example.myapplication.slice;
import com.example.myapplication.ResourceTable;
import ohos.aafwk.ability.AbilitySlice;
import ohos.aafwk.content.Intent;
public class MainAbilitySlice extends AbilitySlice {
    @Override
    public void onStart(Intent intent) {
        super.onStart(intent);
        super.setUIContent(ResourceTable.Layout_ability_main);
        //加载 layout 目录下的 XML 布局
    }
}
```

(6)使用预览器或模拟器运行项目。

2. 代码方式编写第二个页面

使用代码的方式创建第二个页面步骤如下。

(1)在 Project 窗口中,打开 entry→src→main→java→com.example.myapplication,右击 slice 文件夹,选择 New→Java Class,命名为 SecondAbilitySlice,按 Enter 键。

(2)第二个页面上有一个文本。打开 SecondAbilitySlice 文件,添加 Text,相关代码如下。

```java
//根据实际工程/包名引入
package com.example.myapplication.slice;
import ohos.aafwk.ability.AbilitySlice;
import ohos.aafwk.content.Intent;
import ohos.agp.colors.RgbColor;
import ohos.agp.components.DependentLayout;
import ohos.agp.components.Text;
import ohos.agp.components.element.ShapeElement;
import ohos.agp.utils.Color;
import ohos.agp.components.DependentLayout.LayoutConfig;
public class SecondAbilitySlice extends AbilitySlice {
    @Override
    public void onStart(Intent intent) {
        super.onStart(intent);
        //声明布局
        DependentLayout myLayout = new DependentLayout(this);
        //设置布局宽高
        myLayout.setWidth(LayoutConfig.MATCH_PARENT);
        myLayout.setHeight(LayoutConfig.MATCH_PARENT);
        //设置布局背景为白色
        ShapeElement background = new ShapeElement();
        background.setRgbColor(new RgbColor(255, 255, 255));
        myLayout.setBackground(background);
        //创建一个文本
        Text text = new Text(this);
        text.setText("Hi there");
        text.setWidth(LayoutConfig.MATCH_PARENT);
        text.setTextSize(50, Text.TextSizeType.FP);
        text.setTextColor(Color.BLACK);
        //设置文本的布局
        DependentLayout.LayoutConfig textConfig = new DependentLayout.LayoutConfig(LayoutConfig.MATCH_CONTENT, LayoutConfig.MATCH_CONTENT);
        textConfig.addRule(LayoutConfig.CENTER_IN_PARENT);
        text.setLayoutConfig(textConfig);
        myLayout.addComponent(text);
        super.setUIContent(myLayout);
    }
}
```

3. 实现页面跳转

打开第一个页面的 MainAbilitySlice.java 文件，添加按钮的响应逻辑，实现单击按钮跳转到下一页，示例代码如下。

```java
//根据实际工程/包名引入
package com.example.myapplication.slice;
import com.example.myapplication.ResourceTable;
import ohos.aafwk.ability.AbilitySlice;
import ohos.aafwk.content.Intent;
import ohos.agp.components.Button;
public class MainAbilitySlice extends AbilitySlice {
    @Override
    public void onStart(Intent intent) {
        super.onStart(intent);
        super.setUIContent(ResourceTable.Layout_ability_main);
        Button button = (Button) findComponentById(ResourceTable.Id_button);
        //单击按钮跳转至第二个页面
        button.setClickedListener(listener -> present(new SecondAbilitySlice(), new Intent()));
    }
}
```

使用预览器或模拟器运行项目，获得跳转效果。示例工程参考地址为 https://gitee.com/openharmony/app_samples/tree/master/common/HelloWorld，指导开发者通过 Java UI 编写两个简单的页面，实现在第一个页面单击按钮跳转到第二个页面。

2.3.3 使用 JS 语言开发

本节介绍如何编写第一个页面，创建另一个页面和实现页面跳转。

1. 编写第一个页面

编写第一个页面步骤如下。

（1）第一个页面内有一个文本和一个按钮，通过 text 和 button 组件实现。

在 Project 窗口中，选择 entry→src→main→js→default→pages→index，打开 index.hml 文件，添加一个文本和一个按钮，相关代码如下。

```html
<!-- index.hml -->
<div class = "container">
    <!-- 添加一个文本 -->
    <text class = "text">
        Hello World
    </text>
    <!-- 添加一个按钮，按钮样式设置为胶囊型，文本显示为 Next,绑定 launch 事件 -->
    <button class = "button" type = "capsule" value = "Next" onclick = "launch"></button>
</div>
```

(2) 打开 index.css 文件,设置文本和按钮的样式,相关代码如下。

```css
/* index.css */
.container {
    display: flex;
    flex-direction: column;
    justify-content: center;
    align-items: center;
    left: 0px;
    top: 0px;
    width: 100%;
    height: 100%;
}
/* 对 class="text"的组件设置样式 */
.text{
    font-size: 42px;
}
/* 对 class="button"的组件设置样式 */
.button {
    width: 240px;
    height: 60px;
    background-color: #007dff;
    font-size: 30px;
    text-color: white;
    margin-top: 20px;
}
```

(3) 使用预览器或模拟器运行项目,可以看到运行效果。

2. 创建另一个页面

创建另一个页面步骤如下。

(1) 在 Project 窗口中,打开 entry→src→main→js→default,右击 pages 文件夹,选择 New→JS Page,命名为 details,按 Enter 键。创建完成后,可以看到 pages.index 文件夹下的文件目录结构。

(2) 打开 details.hml 文件,添加一个文本,相关代码如下。

```html
<!-- details.hml -->
<div class="container">
    <text class="text">
        Hi there
    </text>
</div>
```

(3) 打开 details.css 文件,设置文本的样式,相关代码如下。

```css
/* details.css */
.container {
```

```css
    display: flex;
    flex-direction: column;
    justify-content: center;
    align-items: center;
    left: 0px;
    top: 0px;
    width: 100%;
    height: 100%;
}
.text {
    font-size: 42px;
    text-align: center;
}
```

3. 实现页面跳转

实现页面跳转步骤如下。

(1) 打开第一个页面的 index.js 文件,导入 router 模块,页面路由 router 根据 URI 找到目标页面,从而实现跳转,相关代码如下。

```js
//index.js
import router from '@system.router';
export default {
  launch() {
    router.push({
      uri:'pages/details/details', //指定要跳转的页面
    })
  }
}
```

(2) 再次使用预览器或模拟器运行项目,达到运行效果。

2.3.4 可视化开发

使用 DevEco Studio V2.2 Beta1 及更高版本。低代码开发功能仅适用于 Phone 设备的 JS 工程,且 compileSdkVersion 必须为 6 及以上。

1. 创建 JS 工程

创建 JS 工程步骤如下。

(1) 打开 DevEco Studio,创建一个新工程,选择支持 Phone 的 Empty Ability 模板,如图 2-12 所示。

(2) 选择 Super Visual,表示使用低代码开发功能开发应用/服务。单击 Finish 按钮等待工程同步完成,如图 2-13 所示。

(3) 同步完成后,工程目录中自动生成低代码目录结构,同时第一个页面创建完毕。其中,pages→index→index.js 是低代码页面的逻辑描述文件,定义页面中所用到的所有逻辑关系,例如数据、事件等。如果创建多个低代码页面,则 pages 目录下会生成多个页面文件

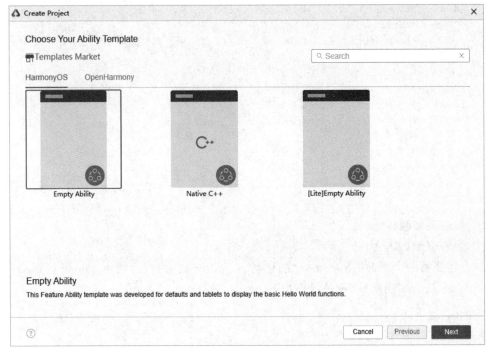

图 2-12　创建新工程

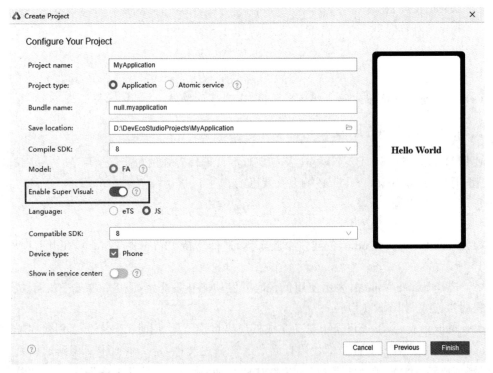

图 2-13　同步完成

夹及对应的JS文件。pages→index→index.visual是visual文件存储低代码页面的数据模型，双击该文件即可打开低代码页面，进行可视化开发设计。如果创建多个低代码页面，则pages目录下会生成多个页面文件夹及对应的visual文件，如图2-14所示。

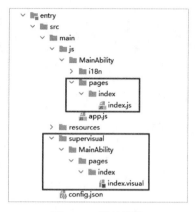

图2-14　目录结构

2. 设置第一个页面

设置第一个页面步骤如下。

（1）第一个页面内有一个文本和一个按钮，通过Text和Button组件实现。

打开index.visual文件，选中画布中的组件，右击选择Delete删除画布原有组件。选中画布，单击右侧属性样式栏中的样式图标（Flex），设置画布的FlexDirection样式为column，使画布的主轴垂直；设置画布的JustifyContent样式为center，使其子组件在主轴上居中显示；设置画布的AlignItems样式为center，使其子组件在交叉轴上居中显示。

（2）选中UI Control中的Text组件，将其拖至中央画布区域。单击右侧属性样式栏中的属性图标（Properties），设置Text组件的Content属性为Hello World；单击右侧属性样式栏中的样式图标（Feature），设置组件的FontSize样式为60px，使其文字放大；设置组件的TextAlign样式为center，使组件文字居中显示。再选中画布上的Text组件，拖动放大。

（3）选中UI Control中的Button组件，将其拖至中央画布区域。单击右侧属性样式栏中的属性图标（Properties），设置Button组件的Value属性为Next；单击右侧属性样式栏中的样式图标（Feature），设置组件的FontSize样式为60px，使其文字放大。

（4）使用预览器或模拟器运行项目，效果如图2-15所示。

3. 创建另一个页面

在Project窗口中，选择工程中的entry→src→main→JS→default→pages，右击选择New→Visual，命名为page2。

第二个页面上有一个文本，通过Text组件实现。打开page2.visual文件，删除画布原有组件。选中画布，设置画布的JustifyContent样式为center，设置画布的AlignItems样式为center。

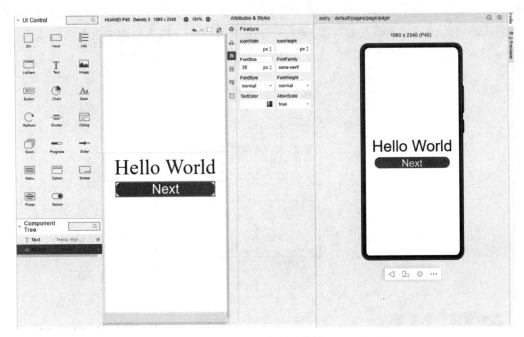

图 2-15　运行项目效果

选中 Text 组件拖至画布，设置 Text 组件的 Content 属性为 Hi there，设置组件的 FontSize 样式为 60px，设置组件的 TextAlign 样式为 center，再选中画布上的 Text 组件，拖动放大。

4．实现页面跳转

（1）在 Project 窗口中，打开工程中的 entry→src→main→JS→default→pages→index→index.js，导入 router 模块，页面路由 router 根据页面的 URI 找到目标页面，从而实现跳转。示例代码如下。

```
import router from '@system.router';
export default {
    launch() {
        router.push({
            uri:'pages/page2/page2', // 指定要跳转的页面
        })
    }
}
```

（2）打开 index.visual，选中画布上的 Button 组件，单击右侧属性样式栏中的事件图标（Events），单击 Click 事件的输入框，选择 launch 事件，操作示例如图 2-16 所示。

（3）使用预览器或模拟器运行项目，达到切换效果。

图 2-16　页面跳转

2.4　DevEco Studio 工程管理

本节详细介绍 DevEco Studio 工程结构的使用方法。HarmonyOS 应用/服务发布形态为 APP Pack(Application Package,APP)，它由一个或多个 HAP 包及描述 APP Pack 属性的 pack.info 文件组成。

一个 HAP 在工程目录中对应一个 Module，它由代码、资源、第三方库及应用/服务配置文件组成，可以分为 Entry 和 Feature 两种类型。Entry 是应用/服务的主模块，可独立安装运行。一个 App 中，对于同一类型的设备，可以包含一个或多个 entry 类型的 HAP，如果同一设备类型包含多个 entry 模块，需要配置 distroFilter 分发规则。Feature 是应用/服务的动态特性模块。一个 App 可以包含一个或多个 feature 类型的 HAP，也可以不含。

HAP 是 Ability 的部署包，HarmonyOS 应用/服务代码围绕 Ability 组件展开，它由一个或多个 Ability 组成。Ability 分为两种类型：FA 和 PA。FA/PA 是应用/服务的基本组成单元，能够实现特定的业务功能。FA 有 UI 界面，而 PA 无 UI 界面，如图 2-17 所示。

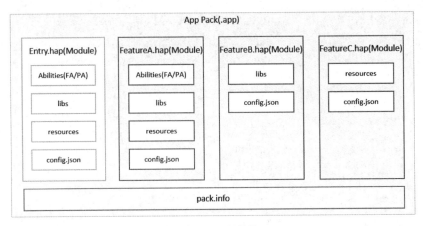

图 2-17 工程结构

2.4.1 工程结构

本节介绍 HarmonyOS App 工程结构,以便快速掌握开发方法。

1. eTS 工程目录结构

eTS 工程目录结构如图 2-18 所示。

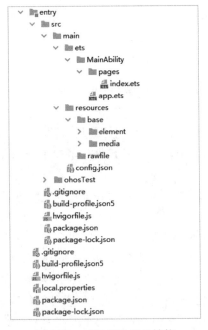

图 2-18 eTS 工程目录结构

entry:HarmonyOS 工程模块,编译构建生成一个 HAP 包。

（1）src→main→eTS：用于存放 eTS 源码。

（2）src→main→eTS→MainAbility：应用/服务的入口。

（3）src→main→eTS→MainAbility→pages：MainAbility 包含的页面。

（4）src→main→eTS→MainAbility→pages→index.ets：pages 列表中的第一个页面，即应用的首页入口。

（5）src→main→eTS→MainAbility→app.ets：承载 Ability 生命周期。

（6）src→main→resources：用于存放应用/服务所用到的资源文件，如图形、多媒体、字符串、布局文件等。

（7）src→main→config.json：模块配置文件。主要包含 HAP 包的配置信息、应用/服务在具体设备上的配置信息以及应用/服务的全局配置信息。

（8）build-profile.json5：当前的模块信息、编译信息配置项，包括 buildOption、targets 配置等。

（9）hvigorfile.js：模块级编译构建任务脚本，开发者可以自定义相关任务和代码实现。

build-profile.json5：应用级配置信息，包括签名、产品配置等。

hvigorfile.js：应用级编译构建任务脚本。

2. Java 工程目录结构

Java 工程目录结构如图 2-19 所示。

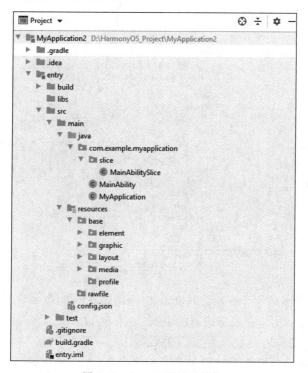

图 2-19　Java 工程目录结构

.gradle：Gradle 配置文件，由系统自动生成，一般情况下不需要修改。

entry：默认启动模块（主模块），开发者用于编写源码文件及开发资源文件的目录。

(1) entry→libs：用于存放 entry 模块的依赖文件。

(2) entry→src→main→Java：用于存放 Java 源码。

(3) entry→src→main→resources：用于存放资源文件，例如图形、多媒体、字符串、布局文件等。

(4) entry→src→main→config.json：应用配置文件。

(5) entry→src→ohos Test：HarmonyOS 应用测试框架，运行在设备模拟器或真机设备上。

(6) entry→src→test：编写代码单元测试的目录，运行在本地 Java 虚拟机（JVM）上。

(7) entry→.gitignore：标识 git 版本管理需要忽略的文件。

(8) entry→build.gradle：entry 模块的编译配置文件。

3. JS 工程目录结构

JS 工程目录结构如图 2-20 所示。

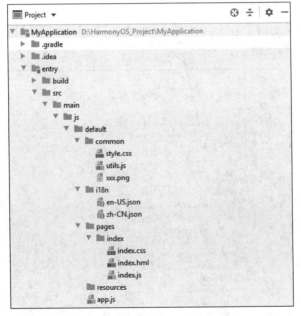

图 2-20　JS 工程目录结构

common 目录（可选）：用于存放公共资源文件，例如媒体资源、自定义组件和 JS 文档等。

i18n 目录（可选）：用于存放多语言的 Json 文件，可以在该目录下定义应用在不同语言系统下显示的内容，如应用文本词条、图片路径等。

pages 目录：pages 文件夹下可以包含 1 个或多个页面，每个页面都需要创建一个文件夹（图 2-20 中的 index）。页面文件夹下主要包含 3 种文件类型，即 CSS、JS 和 HML 文件。

(1) pages→index→index.hml 文件：HML(HarmonyOS Markup Language)是一套类 HTML 的标记语言。页面具备数据绑定、事件绑定、列表渲染、条件渲染和逻辑控制等高级能力。HML 文件定义了页面的布局结构，使用到组件及这些组件的层级关系。

(2) pages→index→index.css 文件：CSS 文件定义了页面的样式与布局，包含样式选择器和各种样式属性等。

(3) pages→index→index.js 文件：JS 文件描述了页面的行为逻辑，此文件中定义了页面里所用到的所有逻辑关系，例如数据、事件等。

resources(可选)：用于存放资源配置文件，例如全局样式、多分辨率加载等配置文件。

app.js 文件：全局的 JavaScript 逻辑文件和应用的生命周期管理。

4. C++工程目录结构

C++工程目录结构如图 2-21 所示。

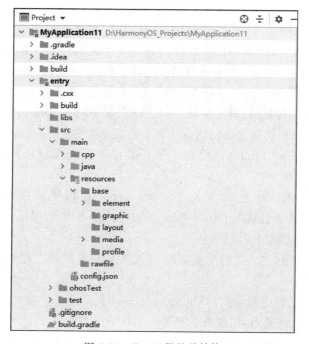

图 2-21 C++工程目录结构

.gradle：Gradle 配置文件，由系统自动生成，一般情况下不需要修改。

entry：默认启动模块(主模块)，开发者用于编写源码文件及开发资源文件的目录。

(1) entry→libs：用于存放 entry 模块的依赖文件。

(2) entry→src→main→cpp：用于存放 C++源码。

(3) entry→src→main→java：用于存放 Java 源码。

(4) entry→src→main→resources：用于存放资源文件，例如图形、多媒体、字符串、布局文件等。

(5) entry→src→main→config.json：应用配置文件。

(6) entry→src→ohos Test：应用测试框架，运行在设备模拟器或真机设备上。

(7) entry→src→test：编写代码单元测试代码的目录，运行在本地 Java 虚拟机(JVM)上。

(8) entry→.gitignore：标识 git 版本管理需要忽略的文件。

(9) entry→build.gradle：entry 模块的编译配置文件。

2.4.2 工程操作

本节介绍 DevEco Studio 的工程操作方法，包括创建新工程，打开现有工程，适配历史工程，导入 Sample 工程，在工程中添加/删除 Module。

1. 创建新工程

开发 HarmonyOS 应用时，根据向导，创建一个新的工程，工具会自动生成对应的代码和资源模板。

(1)通过如下两种方式打开工程创建向导界面。如果当前未打开任何工程，可以在 DevEco Studio 的欢迎页，选择 Create Project 创建一个新工程。如果已经打开工程，可以在菜单栏选择 File→New→New Project 创建一个新工程，如图 2-22 所示。

(2)根据工程创建向导，选择需要的 Ability 工程模板，单击 Next 按钮。DevEco Studio 支持手机、平板、车机、智慧屏、智能穿戴和智慧视觉设备的 HarmonyOS 应用开发，预置了丰富的工程模板，可以根据向导轻松创建适应于各类设备的工程，并自动生成对应的代码和资源模板。同时，DevEco Studio 还提供了多种编程语言进行应用开发，包括 Java、JS 和 C/C++ 编程语言。

单击 Template Market 进入 Template Market 可获取更多模板资源，在 Template Market 选中要下载的工程模板，单击 Download 按钮，下载工程模板后，显示在 Choose your ability template 页面上。

(3)单击 Next 按钮，进入工程配置阶段，需要根据向导配置工程的基本信息，如图 2-23 所示。

Project Name：工程名称，可以自定义。

Project Type：工程类型，标识该工程是一个原子化服务(Service)或传统方式需要安装的应用(Application)。如果是创建的原子化服务，在调试、运行时，设备桌面上没有应用图标，使用 DevEco Studio 的调试和运行功能，启动原子化服务。原子化服务免安装，config.json 中自动添加 installationFree 字段，取值为 True。如果 entry 模块的 installationFree 字段为 True，则其相关的所有 HAP 模块的 installationFree 字段都默认为 True；如果 entry 模块的 installationFree 字段为 False，则其相关的所有 HAP 模块可以配置为 True 或 False。编译构建 App 时，每个 HAP 包大小不能超过 10MB。

Package Name：软件包名称，默认情况下，应用 ID 也会使用该名称，在发布时，ID 需要唯一。

Save Location：工程文件本地存储路径(不能包含中文字符)。

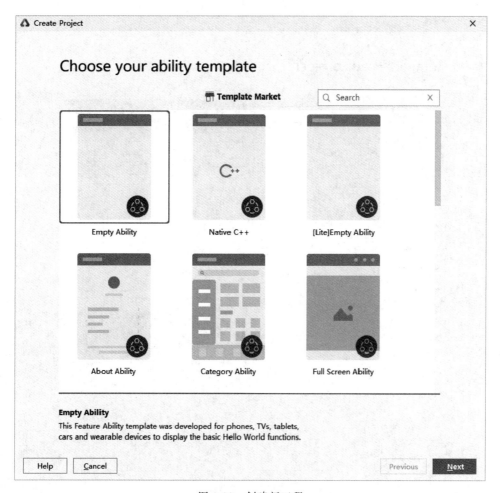

图 2-22 创建新工程

Compatible API Version：兼容的 SDK 最低版本。

Language：该工程模板支持的开发语言。

Device Type：该工程模板设备类型支持多选，默认全部勾选。如果勾选多个设备，表示该原子化服务或传统方式需要安装的应用支持部署在多个设备上。

Show in Service Center：是否在服务中心露出。如果 Project Type 为 Service，则会同步创建一个 2×2 的服务卡片模板，同时还会创建入口卡片；如果 Project Type 为 Application，则只会创建一个 2×2 的服务卡片模板。

Advanced：高级选项，根据选项预览该工程模板的效果图（仅部分工程模板支持效果图预览）。

(4) 单击 Finish 按钮，工具会自动生成示例代码和相关资源，等待工程创建完成。

各模板支持的设备及 API 如表 2-3 所示。

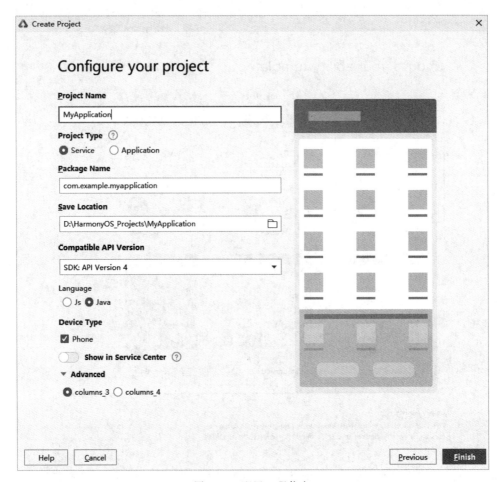

图 2-23 配置工程信息

表 2-3 各模板支持的设备及 API

模板名称	支持设备	开发语言	API 版本	模板说明
Empty Ability	Phone、Tablet、TV、Wearable	JavaScript	4\5\6\7\8	支持低代码开发,用于 Phone、TV、Tablet、Wearable 设备的 Feature Ability 模板,展示 Hello World 基础功能
	Phone、Tablet、TV、Wearable、Car	Java	4\5\6\7\8	用于 Phone、TV、Tablet、Wearable、Car 设备的 Feature Ability 模板,展示 Hello World 基础功能
	Phone	eTS		用于 Phone 设备的 Feature Ability 模板,展示 Hello World 基础功能
Native C++	Phone、Car	C++	4\5\6\7\8	用于 Phone、Car 设备的 Feature Ability 模板,作为 HarmonyOS 应用/服务调用 C++ 代码的示例工程,界面显示 Hello from JNI C++codes

2. 打开现有工程

打开现有工程包括如下两种方式：如果当前未打开任何工程，可以在 DevEco Studio 的欢迎页选择 Open Project；如果已经打开，可以在菜单栏中选择 File→Open 打开现有工程。

3. 导入 Sample 工程

DevEco Studio 支持 HarmonyOS Sample 工程的导入功能，通过对接 Gitee 开源社区中的 Sample 资源，可一键导入 Sample 工程到 DevEco Studio 中。

目前，HarmonyOS 和 OpenHarmony 的 Sample 均在同一个 Gitee 仓中，但 OpenHarmony 的 Sample 示例并不适用于 HarmonyOS SDK。在导入时，请不要导入 OpenHarmony Samples 下的 Sample 中。下面介绍导入 HarmonyOS Sample 的方法。

（1）在 DevEco Studio 的欢迎页，进入 Configure（或图标）→Settings→Version Control→Git 界面，单击 Test 按钮检测是否安装 Git 工具，若已安装，下一步开始导入 Sample，如图 2-24 所示。若未安装，单击 Download and Install，DevEco Studio 会自动下载并安装。安装完成后，开始导入 Sample，如图 2-25 所示。

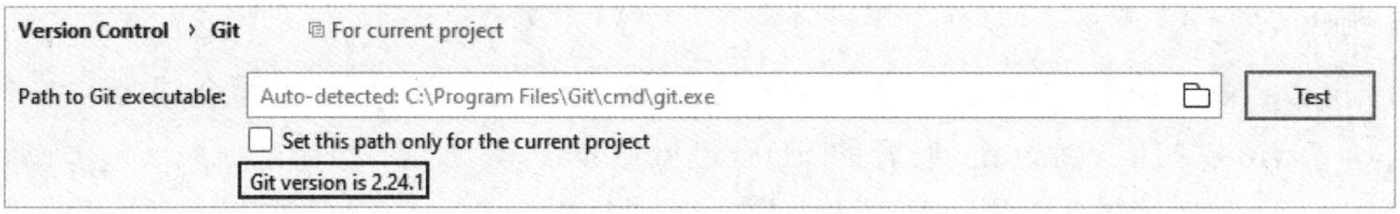

图 2-24　检测安装 Git

图 2-25　导入 Sample

（2）在 DevEco Studio 的欢迎页，单击 Import Sample 按钮，导入 Sample 工程。在打开工程的情况下，可以单击 File→New→Import Sample 进行导入。

（3）在 HarmonyOS Samples 下选择需要导入的 Sample 工程，单击 Next 按钮。

（4）设置 Project Name 和 Project Location，单击 Finish 按钮，等待 Sample 工程导入完成。

（5）导入 Sample 后，等待工程同步完成即可。

4. 在工程中添加/删除 Module

Module 是 HarmonyOS 应用的基本功能单元，包含源代码、资源文件、第三方库及应用配置文件，每个 Module 都可以独立进行编译和运行。一个 HarmonyOS 应用通常会包含一个或多个 Module，因此可以在工程中创建多个 Module，每个 Module 分为 Ability 和 Library（HarmonyOS Library 和 Java Library）两种类型。

在一个 App 中，对于同一类型设备可以包含一个或多个 Entry Module。如果同一类型设备存在多个 Entry Module，还需要配置不同的 Entry 模块的分发规则。

1）创建新的 Module

创建新的 Module 步骤如下。

（1）通过如下两种方法，在工程中添加新的 Module。

方法 1：鼠标指针移到工程目录顶部，右击选择 New→Module，开始创建新的 Module。

方法 2：在菜单栏中选择 File→New→Module，创建新的 Module。

（2）在 New Project Module 界面中，选择需要创建的模板。

（3）单击 Next 按钮，在 Module 配置页面设置新增 Module 的基本信息。Module 类型为 Ability 或者 HarmonyOS Library 时，根据如下内容进行设置。

Application/Library name：新增 Module 所属的类名称。

Module Name：新增模块的名称。

Module Type：仅在 Ability 模板存在，可以选择 Feature 和 Entry 类型。如果同一类型的设备已经存在 Entry 模块，添加新 Module 选择 Entry 类型时，还需要配置 distroFilter 分发规则。

Package Name：软件包名称，可以单击 Edit 按钮修改默认包名称，需全局唯一。

Compatible API Version：兼容的 SDK 最低版本。

Language：选择该 Module 的开发语言。

Device Type：选择模块的设备类型，如果新建模块的 Module Type 为 Feature，则只能选择该工程原有的设备类型；如果 Module Type 为 Entry，可以选择 Module 支持的其他设备类型。

Show in Service Center：是否在服务中心露出，仅在 Ability 模板存在。如果工程类型为 HarmonyOS Service，则会同步创建一个 2×2 的服务卡片模板，同时还会创建入口卡片；如果工程类型为 HarmonyOS Application，则只会创建一个 2×2 的服务卡片模板。

Module 类型为 Java Library 时，根据如下内容进行设置，然后单击 Finish 按钮完成创建。

① Library name：Java Library 类名称。

② Java package name：软件包名称，可以单击 Edit 按钮修改默认包名称，需全局唯一。

③ Java class name：class 文件名称。

④ Create.gitignore file：是否自动创建.gitignore 文件，勾选表示创建。

（4）设置新增 Ability 的 Page Name 和 Layout Name。若 Module 的模板类型为 Ability，还需要设置 Visible 参数，表示 Ability 是否可以被其他应用所调用。勾选（True）可以被其他应用调用，不勾选（False）则不能被其他应用调用。

（5）单击 Finish 按钮，等待创建完成后，可以在工程目录中查看和编辑新增的 Module。

2）导入 Module

HarmonyOS 工程支持从其他工程中导入 HarmonyOS 模块功能，导入的模块只能是

HarmonyOS 工程中的模块；同样，也支持 OpenHarmony 工程导入其他 OpenHarmony 工程的模块。

（1）在主菜单栏中单击 File→New→Import Module。

（2）选择导入的 HarmonyOS 模块，既可以是具体的 HarmonyOS 模块，也可以是 HarmonyOS 工程。

选择 HarmonyOS 模块：如果导入的模块是 Feature 类型，依赖其他 Entry 类型的模块时，会自动勾选其依赖的 Entry 模块。如果依赖的 Entry 模块名与当前工程的模块名冲突，则不会导入。因此，在导入 Feature 模块时，尽量避免其依赖的 Entry 类型的模块名与当前工程模块名重复。

选择 HarmonyOS 工程：会在列表中列出 HarmonyOS 工程下的所有模块。如果选择的类型为 Feature 模块，则会自动勾选其依赖的 Entry 模块。

（3）单击 Finish 按钮，等待完成导入。

3）配置 distroFilter 分发规则

同一类型的设备（Phone、Wearable、Lite Wearable 等）可能在系统 API 版本（apiVersion）、屏幕形状（screenShape）、分辨率（screenWindow）存在差异。针对这些差异，需要对同一类型设备的不同型号进行适配开发，然后在应用市场实现精准分发，以便不同设备的消费者用户能获得更好的体验。为实现精准分发，需要在一个工程中，针对同一类型设备，添加多个 Entry 模块适配不同型号，再配置不同的分发规则。

因此，在同一个工程中，同一个设备存在多个 Entry 模块，需要在每个 Entry 模块的 config.json 文件中配置 distroFilter 分发规则，具体规则如下。

通过 DeviceType 与 apiVersion、screenShape、screenWindow 的组合确定唯一一个 Entry。

distroFilter 中至少包含 apiVersion、screenShape 和 screenWindow 中的一个标签。

一般情况下，apiVersion 标签用于 Phone、Tablet、TV、Wearable、Car 和 Lite Wearable 的设备中；screenShape 和 screenWindow 标签用于 Lite Wearable 设备中。

如果一个 Entry 模块中配置了 apiVersion、screenShape 和 screenWindow 中的任意一个或多个标签，则其他的 Entry 模块也必须包含相同的标签。

apiVersion、screenShape 和 screenWindow 标签的配置格式如下。

```
"module": {
  ...
  "distroFilter": {
    "标签名字": {
      "policy": "include|exclude"
      "value": [ a, b, c]
    }
  }
}
```

其中，policy 取值为 include 时，表示设备满足 value 取值，应用市场向该设备进行分发；policy 取值为 exclude 时，表示除 value 的取值外其他合法的取值，应用市场都会向设备进行分发。screenWindow 标签的 policy 取值只能为 include。

4）删除 Module

为防止开发者在删除 Module 的过程中误将其他的模块删除，DevEco Studio 提供统一的模块管理功能，需要先在模块管理中移除对应的模块后，才允许删除。

在菜单栏中选择 File→Project Structure→Modules，选择需要删除的 Module，单击"-"按钮，并在弹出的对话框中单击 Yes 按钮。

在工程目录中选中该模块，右击选中 Delete，并在弹出的对话框中单击 Delete 按钮。

2.4.3 HarmonyOS 共享包

HarmonyOS 共享包又称为 HAR（HarmonyOS Ability Resources）包，可以提供应用/服务构建所需的一切内容，包括源码、资源文件、HarmonyOS 配置文件及第三方库。HAR 不同于 HAP，HAR 不能独立安装运行在设备上，只能作为应用/服务模块的依赖项被引用。该功能适用于 HarmonyOS API 4-7 的工程。HAR 包只能被 Phone、Tablet、Car、TV 和 Wearable 工程所引用。

1. 创建库模块

在 DevEco Studio 中，可以通过如下方式创建新的库模块。

（1）在工程中添加新的 HarmonyOS Library 模块。

方法 1：鼠标指针移到工程目录顶部，右击选择 New→Module。

方法 2：在菜单栏中选择 File→New→Module。

（2）在 New Module 界面中，选择 HarmonyOS Library，并单击 Next 按钮。

（3）在 Configure the New Module 界面中，设置新添加的模块信息如下，单击 Finish 按钮完成创建。

Application/Library name：新增 Module 所属的类名称。

Module Name：新增模块的名称。

Package Name：软件包名称，可以单击 Edit 按钮修改默认包名称，需全局唯一。

Compatible API Version：兼容的 SDK 版本。

Device Type：选择设备类型，支持选择多设备。

（4）等待工程自动同步完成后，会在工程目录中生成对应的库模块。

2. 将库模块编译为 HAR

利用 Gradle 可以将 HarmonyOS Library 库模块构建为 HAR 包，以便在工程中引用 HAR 或者将 HAR 包提供给其他开发者调用，构建方法如下。

在 Gradle 构建任务中，双击 PackageDebugHar 或 PackageReleaseHar 任务，构建 Debug 类型或 Release 类型的 HAR。编译 debug 类型 HAR 包如图 2-26 所示，编译 Release 类型 HAR 包如图 2-27 所示。

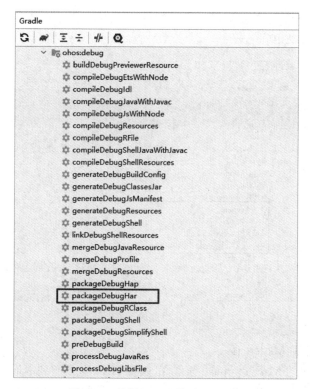

图 2-26　编译 debug 类型 HAR 包

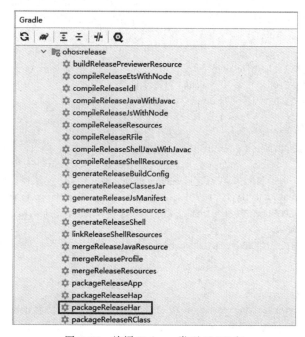

图 2-27　编译 Release 类型 HAR 包

待构建任务完成后，可以在工程目录中的 moduleName→build→outputs→har 获取生成的 HAR 包，如图 2-28 所示。

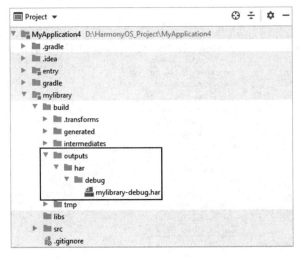

图 2-28　生成的 HAR 包

3．发布 HAR 包到 Maven 仓

借助 Gradle 提供的 Maven-publish 插件，可以将 HAR 包发布到本地或远程 Maven 仓。

(1) 在工程根目录下，右击 New→File，创建以 .gradle 结尾的文件，例如 upload.gradle。

(2) 在创建 upload.gradle 文件中，如下示例代码为发布 HAR 包到 Maven 仓的最小集，请根据实际发布信息进行修改。

```
apply plugin: 'maven-publish'
def DEFAULT_POM_NAME = 'myLibrary'
def DEFAULT_POM_VERSION = '1.0.1'    //HAR 包版本信息
def DEFAULT_POM_ARTIFACT_ID = "harTest"    //HAR 包 ID
def DEFAULT_POM_GROUP_ID = 'com.huawei.har'    //项目组 ID
def DEFAULT_POM_PACKAGING = 'har'    //包类型,固定为 HAR
def DEFAULT_POM_DESCRIPTION = 'myLib for harmonyos'
def MAVEN_USERNAME = 'admin'    //远程 Maven 仓的用户名
def MAVEN_PASSWORD = '******'    //远程 Maven 仓的密码
def LOCAL_MAVEN_REPOSITORY_URL = 'D:/01.localMaven/'    //本地 Maven 仓地址
def REMOTE_MAVEN_REPOSITORY_URL = 'https:// '    //远程 Maven 仓地址
afterEvaluate { project ->
    DEFAULT_POM_ARTIFACT_ID = project.name
    publishing {
        publications {
            maven(MavenPublication) {
                from components.debug    //指定发布的 HAR 包类型为 debug 或 release
                group = DEFAULT_POM_GROUP_ID
                artifactId = DEFAULT_POM_ARTIFACT_ID
```

```
                version = DEFAULT_POM_VERSION
                pom {
                    name = DEFAULT_POM_NAME
                    description = DEFAULT_POM_DESCRIPTION
                    packaging = DEFAULT_POM_PACKAGING
                }
            }
        }
        repositories {
            maven {
                url = LOCAL_MAVEN_REPOSITORY_URL   //发布到本地 Maven 仓
                //发布到远程 Maven 仓的地址及 Maven 仓的账号和密码
                / * url = REMOTE_MAVEN_REPOSITORY_URL
                 credentials {
                    username MAVEN_USERNAME
                    password MAVEN_PASSWORD
                 }
                 * /
            }
        }
    }
}
```

（3）在 HAR 模块的 build.gradle 中，添加 HAR 发布脚本。添加完成后，单击 Sync Now 进行同步，如图 2-29 所示。

```
apply from:'../upload.gradle'
```

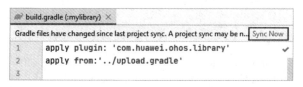

图 2-29　添加脚本

（4）同步完成后，会在 Gradle 任务中增加 publishing 的任务列表，如图 2-30 所示。

（5）双击执行 publishMavenPublicationToMavenRepository 任务，将 HAR 包发布到指定的 Maven 地址。

（6）本示例以发布到本地 Maven 地址为例，如图 2-31 所示。

4．为应用模块添加依赖

在应用模块(entry 或 feature)中调用 HAR，常用的添加依赖方式包括如下 3 种。

（1）调用同一个工程中的 HAR：HAR 包和应用模块在同一个工程，打开应用模块的 build.gradle 文件，在 dependencies 闭包中，添加如下代码。添加完成后，单击 Sync Now 同步工程。

图 2-30　任务列表

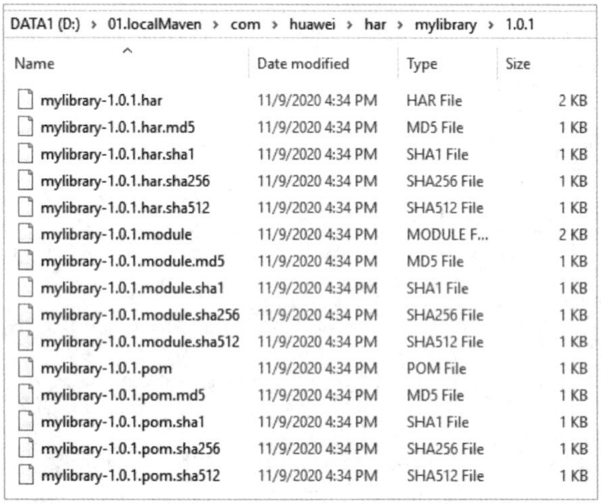

图 2-31　发布结果

```
dependencies {
    implementation project(":mylibrary")
}
```

（2）调用 Maven 仓中的 HAR：无论 HAR 包是本地 Maven 仓还是远程 Maven 仓，均可以采用如下方式添加依赖。在工程 build.gradle 的 allprojects 闭包中，添加 HAR 所在的 Maven 仓地址。

```
repositories {
    maven {
        url 'file://D:/01.localMaven/'    //添加Maven仓地址,可以是本地或远程
    }
}
```

在应用模块 build.gradle 的 dependencies 闭包中添加如下代码。

```
dependencies {
    implementation 'com.huawei.har:mylibrary:1.0.1'
}
```

添加完成后,单击 Sync Now 同步工程。

(3) 调用本地 HAR：将 HAR 包放到模块下的 libs 目录下,检查 build.gradle 中是否添加了 *.har 的依赖。

```
dependencies {
    ...
    implementation fileTree(dir: 'libs', include: ['*.jar', '*.har'])
}
```

2.5 DevEco Studio 开发方法

本节介绍 DevEco Studio 应用开发的基本方法,包括低代码开发、添加 Ability、添加 JS Component 和 JS Page、跨设备代码编辑、定义 HarmonyOS IDL 接口、服务卡片操作、使用预览器查看应用效果、SVG 文件转换为 XML 文件。

2.5.1 低代码开发

HarmonyOS 低代码开发方式,具有丰富的 UI 编辑功能,遵循 HarmonyOS JS 开发规范,通过可视化界面开发方式快速构建布局,可有效降低用户的时间成本并提升用户构建 UI 的效率。

1. 使用低代码开发界面

低代码开发功能适用于 Phone 和 Tablet 设备的 HarmonyOS 应用、原子化服务,目前支持 eTS/JS 语言,且 compileSdkVersion 必须为 6,下面以创建一个新工程为例进行说明。

(1) 打开 DevEco Studio,创建一个新工程,选择支持 Phone 的模板,例如 Empty Ability。

(2) 在工程配置向导中,Device Type 选择 Phone,其他参数根据实际需要设置即可。选择 Super Visual,表示使用低代码开发功能开发应用/服务。单击 Finish 按钮等待工程同步完成,如图 2-32 所示。

(3) 如果从已有工程中创建,选中模块的 pages 文件夹,选择 New→JS Visual,如图 2-33 所示。

图 2-32　低代码开发创建模板

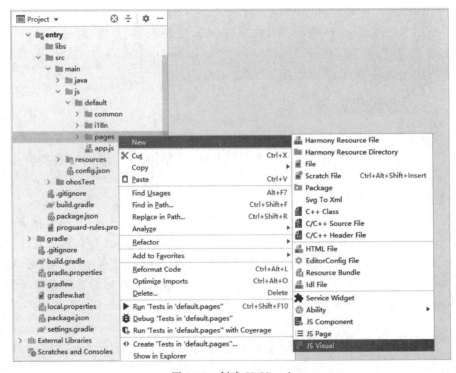

图 2-33　创建 JS Visual

（4）在弹出的对话框中，输入 JS Visual Name，单击 Finish 按钮。创建 JS Visual 后，会在工程中自动生成低代码的目录结构，如图 2-34 所示。

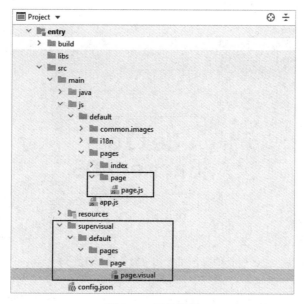

图 2-34　目录结构

pages→page→page.js：低代码页面的逻辑描述文件，定义页面中用到的所有逻辑关系，例如数据、事件等。如果创建多个低代码页面，则 pages 目录下会生成多个页面文件夹及对应的 JS 文件。使用低代码页面开发时，其关联 JS 文件的同级目录中不能包含 HML 和 CSS 页面，例如 JS→default→pages→page 目录下不能包含 HML 与 CSS 文件，否则会出现编译报错。

pages→page→page.visual：visual 文件存储低代码页面的数据模型，双击该文件即可打开低代码页面，进行可视化开发设计。如果创建多个低代码页面，则 pages 目录下会生成多个页面文件夹及对应的 visual 文件。DevEco Studio 预置了 JS Visual 模板，该模板对应的 SDK 版本为 API 7。因此，在创建 JS Visual 文件时，如果模块的 compileSdkVersion 低于 7，则会对新建的 JS Visual 文件对应的 SDK 版本进行降级处理，使其与模块对应的 SDK 版本保持一致。不建议通过文本编辑的方式更改 visual 文件，否则可能导致无法正常使用低代码功能。

（5）打开 page.visual 文件即可进行页面的可视化布局设计与开发，如图 2-35 所示。

使用低代码开发界面过程中，如果界面需要使用到其他暂不支持可视化布局的控件，可以在低代码界面开发完成后，单击"完成"按钮，将低代码界面转换为 HML 和 CSS 代码。注意：代码转换操作会删除 visual 文件及其父目录，且为不可逆过程，代码转换后不能通过 HML/CSS 反向生成 visual 文件。多设备开发的场景，可以单击界面画布右上角设备/模式切换按钮，进行设备切换或模式切换。

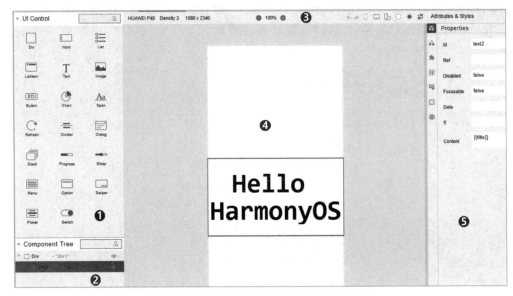

图 2-35 可视化布局设计与开发

(1) UI Control：UI 控件栏，可以将相应的组件选中并拖动到画布(Canvas)中，实现控件的添加。

(2) Component Tree：组件树，在低代码开发界面中，可以直观地看到组件的层级结构、摘要信息及错误提示。通过选中组件树中的组件(画布中对应的组件被同步选中)，实现画布内组件的快速定位；单击图标，可以隐藏/显示相应的组件。

(3) Panel：功能面板，包括常用的画布缩小放大、撤销、显示/隐藏组件虚拟边框、设备切换、模式切换、可视化布局界面一键转换为 HML 和 CSS 文件等。

(4) Canvas：画布，开发者可在此区域对组件进行拖曳、拉伸等可视化操作，构建 UI 布局效果。

(5) Attributes & Styles：属性样式栏，选中画布中的相应组件后，在右侧属性样式栏可以对该组件的属性样式进行配置，具体内容如下。

Properties：对应图标，设置组件基本标识和外观显示特征的属性，例如组件的 ID、If 等属性。

General：对应图标，设置 Width、Height、Background、Position、Display 等常规样式。

Feature：对应图标，设置组件特有样式，例如描述 Text 组件文字大小的 FontSize 样式等。

Flex：对应图标，用于设置 Flex 布局相关样式。

Events：对应图标，为组件绑定相关事件，并设置绑定事件的回调函数。

Dimension：对应图标，用于设置 Padding、Border、Margin 等与盒式模型相关的样式。

Grid：对应图标，用于设置 Grid 网格布局相关样式，该图标只有 Div 组件的 Display 样式被设置为 grid 时才会出现。

Atom：对应图标，用于设置原子布局相关样式。

2. 低代码多语言支持

低代码页面支持多语言能力，使开发者无须开发多个不同语言的版本，通过定义资源文件和引用资源两个步骤使用多语言能力。

（1）在指定的 i18n 文件夹内创建多语言资源文件及对应字符串信息，如图 2-36 所示。

图 2-36　创建多语言文件

（2）在低代码页面的属性样式栏中使用 $t 方法引用资源，系统将根据当前语言环境和指定的资源路径（通过 $t 的 path 参数设置），显示对应语言资源文件中的内容。如图 2-37 所示，在属性栏中引用了字符串资源后，打开预览器即可预览展示效果。

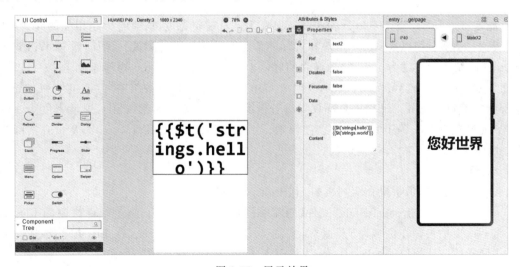

图 2-37　展示效果

3. 低代码屏幕适配

对于屏幕适配问题，低代码页面支持两种配置方法，如图 2-38 所示。

图 2-38　配置方法

指定 designWidth 为 720px，designWidth 为屏幕逻辑宽度，所有与大小相关的样式（如 Width、FontSize）均以 designWidth 和实际屏幕宽度的比例进行缩放。例如，设置 Width 为 100px 时，在实际宽度为 1440 物理像素的屏幕上，Width 实际为 200 物理像素。

设置 autoDesignWidth 为 true，此时 designWidth 字段将会被忽略，渲染组件和布局时按屏幕密度进行缩放。低代码页面仅支持分辨率 1080×2340 像素(P40)，屏幕密度为 3 的场景，此场景下 1px 等于渲染出 3 物理像素。例如，设置 Width 为 100px 时，Width 实际为 300 物理像素。

4. 低代码开发

本示例完成华为手机介绍列表的 HarmonyOS 应用示例。

(1) 根据打开的 page.visual 文件，删除模板页面中的控件后，选中组件栏中的 List 组件，将其拖至中央画布区域，松开鼠标，实现一个 List 组件的添加。在 List 组件添加完成后，用同样的方法拖曳一个 ListItem 组件至 List 组件内，如图 2-39 所示。

(2) 选中画布内的 List 组件，按住控件的 resize 按钮，将 List 拉大，如图 2-40 所示。

(3) 依次选中 Image、Div、Text 组件，将 Image、Div 组件拖至中央画布区域的 ListItem 组件内，Text 组件拖至 Div 组件内。

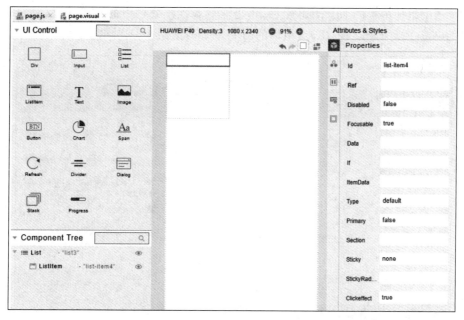

图 2-39 添加组件

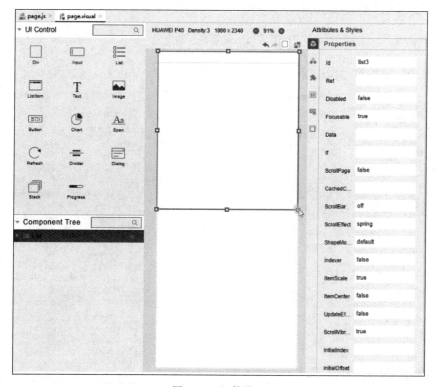

图 2-40 组件设置

(4) 选中组件树中的 ListItem 组件，单击右侧属性样式栏中的样式图标(General)，在展开的 General 栏中修改 ListItem 组件的高度为 100，如图 2-41 所示。

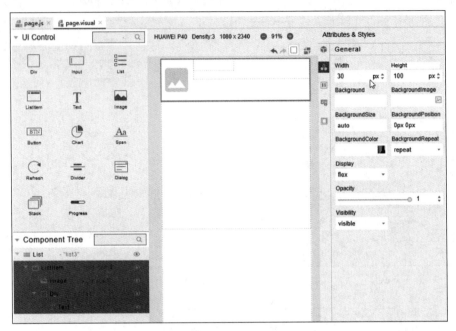

图 2-41　设置组件高度

(5) 对 Div 组件的样式进行调整，如图 2-42 所示。

图 2-42　调整样式

(6) JS 的数据及方法关联。

JS 文件用来定义页面的业务逻辑,基于 JavaScript 语言的动态化能力,可以使应用更加富有表现力,具备灵活的设计。低代码页面支持设置属性(Properties)和绑定事件(Events)时关联 JS 文件中的数据及方法如下,如图 2-43 所示。

```
export default {
    data: {
        title: "Hello HarmonyOS",
        isHarmonyOS: true,
        phoneList: [{
                    title: 'HUAWEI Mate 40 Pro',
                    subTitle: 'Leap Further Ahead',
                    img: 'common/images/Mate40.png',
        }, {
                    title: 'HUAWEI Mate 40 RS Porsche Design',
                    subTitle: 'Pays Tribute To Old World',
                    img: 'common/images/Mate40RS.png',
        }, {
                    title: 'HUAWEI Mate 30',
                    subTitle: 'Rethink Possibilities',
                    img: 'common/images/Mate30.png',
        }, {
                    title: 'HUAWEI Mate 30 5G',
                    subTitle: 'Rethink Possibilities',
                    img: 'common/images/Mate305G.png',
        }],
    },
```

图 2-43　定义数组

(1) 在低代码页面关联 JS 文件的 data 对象中定义 phoneList 数组。

(2) 选中组件树中的 ListItem 组件,单击右侧属性样式栏中的属性图标(Properties),在展开的 Properties 栏中单击 ItemData 属性对应的输入框,并在弹出的下拉框中选择{{phoneList}},实现在低代码页面内引用关联 JS 文件中定义的数据。成功实现关联后,ItemData 属性会根据设置的数据列表(phoneList),展开当前元素,即复制出 3 个结构一致的 ListItem,如图 2-44 所示。

(3) 选中画布中的 Image 组件,修改右侧属性栏中的 Src 属性为{{$item.img}},为 Image 设置图片资源。其中,item 为 phoneList 数组中定义的对象,item.img 为对象中的 img 属性。单击 Feature 图标,修改 ObjectFit 为 contain,如图 2-45 所示。

(4) 选中画布中的 Text 组件,修改右侧属性栏中的 Content 属性为{{$item.title}},为 Text 设置文本内容,单击 Feature 图标调整 Text 的 Width 和 FontSize 样式。

(5) 复制并粘贴画布中的 Text 组件,修改被粘贴出来的 Text 组件右侧属性栏中的 Content 属性为{{$item.subTitle}},为其设置文本内容,单击 Feature 图标调整 Text 的 FontSize 样式,如图 2-46 所示。

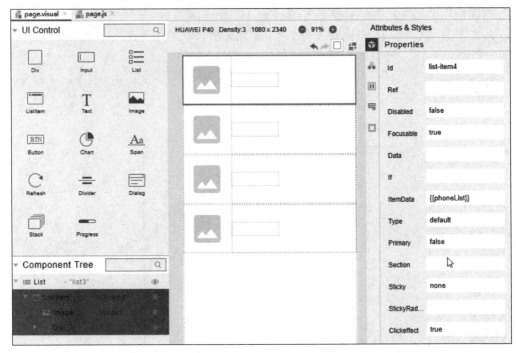

图 2-44 关联数据

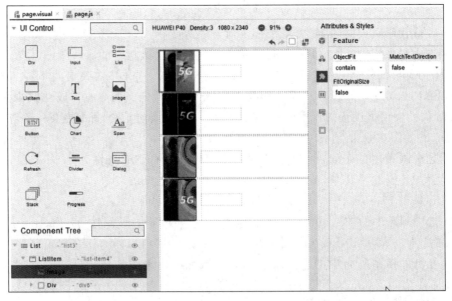

图 2-45 图片设置

图 2-46　调整 FontSize 样式

（6）修改关联 JS 文件中的 switchTitle() 方法，单击 Events 图标，为粘贴出来的 Text 组件绑定 Click 事件，如图 2-47 所示，关联 JS 文件中的 switchTitle 方法。关联后，在 previewer、模拟器及真机中单击该 Text 组件，会将文本内容从 Leap Further Ahead 切换成 Kirin 9000。

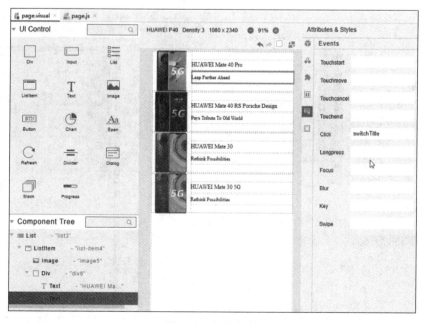

图 2-47　切换文件

page.js 页面如图 2-48 所示。

```js
export default {
    data: {
        title: "Hello HarmonyOS",
        isHarmonyOS: true,
        phoneList: [{
                    title: 'HUAWEI Mate 40 Pro',
                    subTitle: 'Leap Further Ahead',
                    img: 'common/images/Mate40.png',
        }, {
                    title: 'HUAWEI Mate 40 RS Porsche Design',
                    subTitle: 'Pays Tribute To Old World',
                    img: 'common/images/Mate40RS.png',
        }, {
                    title: 'HUAWEI Mate 30',
                    subTitle: 'Rethink Possibilities',
                    img: 'common/images/Mate30.png',
        }, {
                    title: 'HUAWEI Mate 30 5G',
                    subTitle: 'Rethink Possibilities',
                    img: 'common/images/Mate30SG.png',
        }],
    },
    switchTitle() {
        let that = this;
        that.phoneList[0].subTitle =
            (that.phoneList[0].subTitle === "Leap Further Ahead") ? "Kirin 9000" : "Leap Further Ahead";
    }
}
```

图 2-48 page.js 页面

（7）使用预览器预览界面效果。打开 .visual 文件，单击 DevEco 右侧的 Previewer，即可实现实时预览功能，在低代码页面中的每步操作都会通过 Previewer 进行实时显示，如图 2-49 所示。

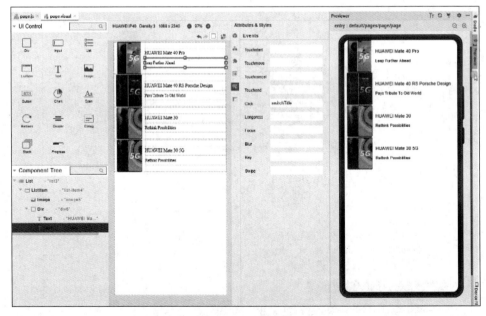

图 2-49 预览功能

2.5.2 添加 Ability

Ability 是应用所具备的能力抽象，一个 Module 可以包含一个或多个 Ability。Ability 分为两种类型：FA(Feature Ability)和 PA(Particle Ability)，DevEco Studio 支持创建的 Ability 模板和应用场景如表 2-4 所示。

表 2-4 Ability 说明

类型	Ability 模板	支持的设备	使 用 场 景
Particle Ability	Empty Data Ability	Phone、Tablet、Car、TV、Wearable	Data Ability 有助于应用管理自身和其他应用所存储数据的访问，并提供与其他应用共享数据的方法。Data 既可用于同设备不同应用的数据共享，也支持跨设备之间不同应用的数据共享
	Empty Service Ability	Phone、Tablet、Car、TV、Wearable	Service Ability 可在后台长时间运行而不提供用户交互界面。Service 可由其他应用或 Ability 启动，即使用户切换到其他应用，Service 仍将在后台继续运行
Feature Ability	Empty Page Ability(JS)	Phone、Tablet、TV、Wearable	用 JS 编写带 UI 的空模板
	Empty Page Ability(Java)	Phone、Tablet、Car、TV、Wearable	用 Java 编写带 UI 的空模板

1. 创建 Particle Ability

创建 Particle Ability 步骤如下。

(1) 选中对应的模块，右击选择 New→Ability，然后选择 Empty Data Ability 或者 Empty Service Ability，如图 2-50 所示。下面以 Wearable 设备为例，不同设备支持的 Ability 模板不同。

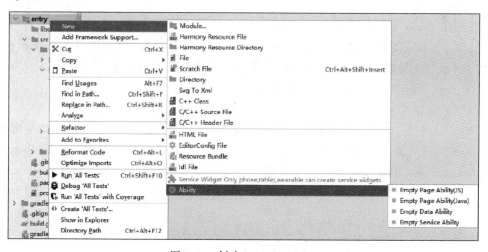

图 2-50 创建 Particle Ability

（2）根据选择的 Ability 模板，设置基本信息。

Empty Data Ability 设置基本信息如下。

① Data Ability Name：Data Ability 类名称。

② Package Name：新增 Ability 对应的包名称。

Empty Service Ability 设置基本信息如下。

① Service Ability Name：Service Ability 类名称。

② Package Name：新增 Ability 对应的包名称。

③ Enable background mode：指定用于满足特定类型的后台服务，可以将多个后台服务类型分配给特定服务，如表 2-5 所示。

表 2-5　各服务与 config.json 文件的映射关系

Background modes	对应 config.json 字段名称	功　能　描　述
Data upload/download, backup/restore	dataTransfer	通过网络/对应设备进行数据下载、备份分享、传输等业务
Audio playback	audioPlayback	音频输出业务
Audio recording	audioRecording	音频输入业务
Picture-in-picture	pictureInPicture	画中画，小窗口播放视频业务
Voice/video call over IP	voip	音视频电话、VOIP 业务
Location update	location	定位、导航业务
Bluetooth communication	bluetoothInteraction	蓝牙扫描、连接、传输业务（穿戴）
Wifi Communication	wifiInteraction	WLAN 扫描、连接、传输业务（多屏，克隆）
Multi-Device connection	multiDeviceConnection	多设备互联业务

（3）单击 Finish 按钮完成 Ability 的创建，可以在工程目录对应的模块中查看和编辑 Ability。

2．创建 Feature Ability

创建 Feature Ability 步骤如下。

（1）选中对应模块，右击选择 New→Ability，然后选择对应的 Feature Ability 模板，如图 2-51 所示。下面以 Wearable 设备为例，不同设备支持的 Ability 模板不同。

（2）根据选择的 Ability 模板，设置 Feature Ability 基本信息如下。

Page Ability Name：Feature Ability 类名称。

Launcher Ability：表示 Ability 在终端桌面上是否启动图标，一个 HAP 可以有多个启动图标，启动不同的 FA。创建 JS Feature Ability 时，会创建一个新的 JS Component。如果勾选 Launcher Ability，表示 Ability 会作为启动页，则需要在 mainAbility.java 中的 onStart 函数中添加如下代码。

```
public void onStart(Intent intent) {
    setInstanceName("default2");    //default2 为创建的 JS Component 名称
    ...
}
```

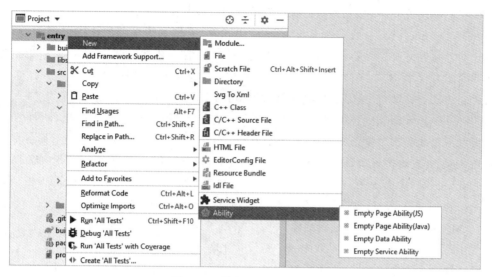

图 2-51　创建 Feature Ability

JS Component Name：JS 组件名称，只有涉及 JS 开发语言时才需要设置。

Package Name：新增 Ability 对应的包名称。

（3）单击 Finish 按钮完成 Ability 的创建，可以在工程目录对应的模块中查看和编辑 Ability。

2.5.3　添加 JS Component 和 JS Page

在支持 JS 语言的工程中，支持添加新的 JS Component 和 JS Page。

JS Component：在 JS 工程中，可以存在多个 JS Component（JS 目录下的 default 文件夹就是一个 JS Component），一个 JS FA 对应一个 JS Component，可以独立编译、运行和调试。Lite Wearable 和 Smart Vision 设备对应的 JS 工程，只存在一个 JS FA，因此 Lite Wearable 和 Smart Vision 设备的 JS 工程不允许创建新的 JS Component。

JS Page：Page 是表示 JS FA 的一个前台页面，由 JS、HML 和 CSS 文件组成，是 Component 的最基本单元，构成 JS FA 的每个界面。

1. 添加 JS Component

在 JS 工程目录中，选中 JS 文件夹，右击选择 New→JS Component，输入 JS Component Name，单击 Finish 按钮完成添加。

2. 添加 JS Page

在 JS 工程目录中，选择需要添加 Page 的 Component 下的 pages 文件夹，右击选择 New→JS Page，输入 JS Page Name，单击 Finish 按钮完成添加。

2.5.4　跨设备代码编辑

HarmonyOS 应用支持在 Phone、TV 和 Wearable 等设备上运行，为适配同一个应用可以运行在多种设备上,开发者需要针对各类设备进行开发和适配。

为此,DevEco Studio 面向跨设备的 JS 应用开发,提供了跨设备的代码编辑能力,可以帮助开发者高效完成代码的开发,减少代码的复杂度。在编辑代码时,可以自动联想出各设备支持的能力合集,例如 HML 组件合集、CSS 样式合集、JS API 接口合集等,开发者可以根据对应的设备类型,快速完成代码的编写和补齐。

同时,DevEco Studio 还提供了兼容性检测功能,可以检测出被调用的组件、样式或者 API 接口是否能够在多设备中兼容,如果存在无法兼容的情况,会提示该代码不具备多设备兼容性,提醒开发者进行确认。如果存在不兼容所有设备的组件、样式或者 API 接口,经过确认后,已在代码中实现相应的逻辑处理,也可以忽略该提示信息。跨设备代码编辑功能在 DevEco Studio V2.0.12.201 及后续版本支持,Java 语言暂不支持。

1. 如何使用跨设备代码编辑功能

使用跨设备代码编辑,需要创建一个跨设备工程,方法如下。

(1) 创建一个新工程:选择任意支持跨设备的 JS Ability 模板,本示例以创建一个 Empty Ability(JS)工程为例。在配置工程信息界面,Device Type 勾选多个设备,如图 2-52 所示。

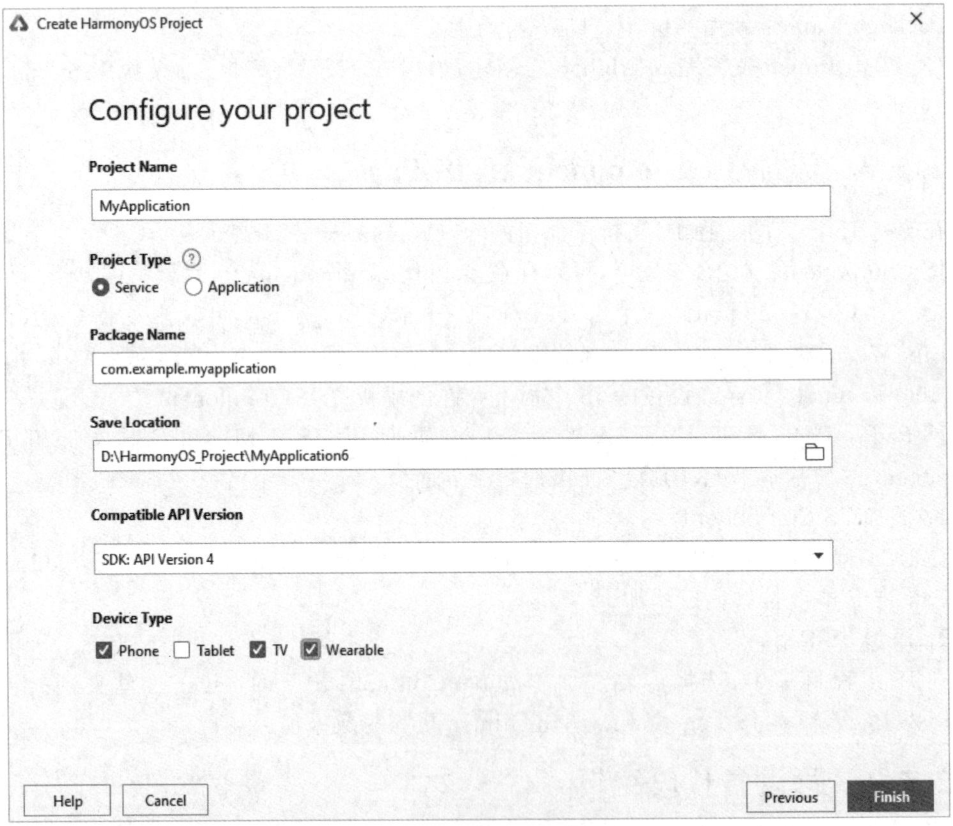

图 2-52 创建 Empty Ability(JS)工程

(2) 打开一个已有工程：打开工程目录模块名(entry)→src→main 下的 config.json 文件，在 module 闭包的 deviceType 中，根据实际支持的设备情况，增加设备类型，例如 TV、Wearable 和 Phone 等，然后重新同步工程，如图 2-53 所示。

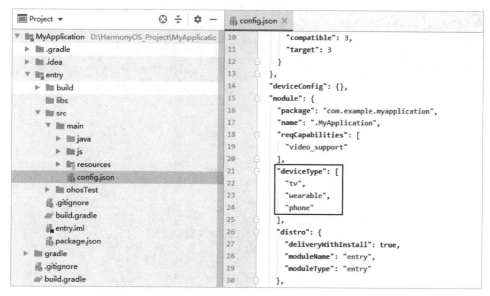

图 2-53　增加设备类型

2．跨设备代码校验

DevEco Studio 支持对 HML、CSS 和 JS 源代码进行检查，根据 config.json 中设置的支持设备类型列表，检查源代码中引用的组件、样式或 API 接口是否与设备类型相匹配，开发者可以根据实际支持的设备类型进行确认。

(1) 单击 File→Settings 或按快捷键 Ctrl＋Alt＋S(Mac 系统为 Configure→Preferences，快捷键 Command＋)，输入 inspect，确保 CSS、HML 和 JavaScript 选项已勾选，如图 2-54 所示。

(2) 单击菜单 Analyze→Inspect Code 设置检查的范围，例如整个工程、某个模块或者具体文件，单击 OK 按钮执行兼容性检查，如图 2-55 所示。

(3) 检查完成后，会在 Inspection Results 中输出检查结果。如图 2-56 所示，该模块支持的设备类型包括 TV、Wearable 和 Phone，但是 picker 只支持 TV 和 Phone。同样在代码编辑器中，如果存在不兼容的字段，编辑器也会提示，可以将鼠标指针放在提示的字段上查看具体提示信息。

2.5.5　定义 HarmonyOS IDL 接口

HarmonyOS Interface Definition Language(简称 HarmonyOS IDL)是 HarmonyOS 的接口描述语言。HarmonyOS IDL 与其他接口语言类似，通过定义客户端与服务器端均认

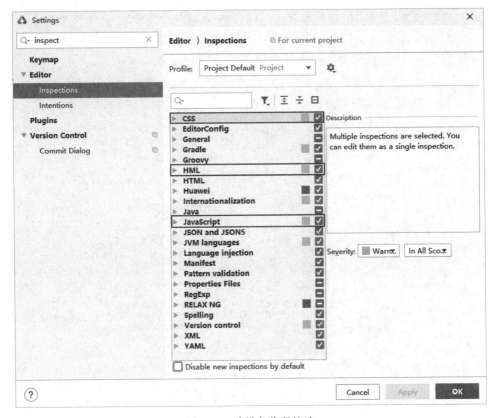

图 2-54　跨设备代码校验

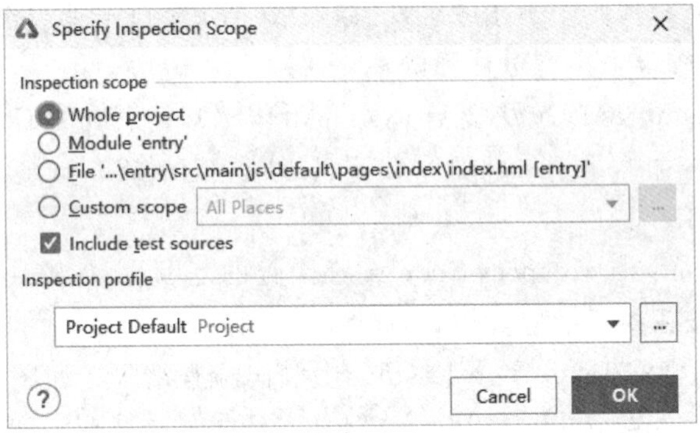

图 2-55　兼容性检查

可的编程接口,可以实现在二者间的跨进程通信(Inter-Process Communication,IPC)。跨进程通信意味着可以在一个进程访问另一个进程的数据,或调用另一个进程的方法。

通常把应用接口提供方(供调用)称为服务器端,调用方称为客户端。客户端通过绑定服务器端的 Ability 与之进行交互,类似于绑定服务。注:只能使用 Java 或 C++语言构建.idl 文件,因此仅 Java、Java+JS 和 C/C++工程支持 IDL。创建.idl 文件步骤如下。

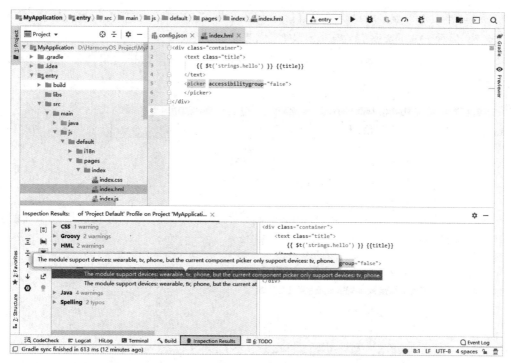

图 2-56　检查结果

（1）在已经创建/打开的 HarmonyOS 工程中，选择 module 目录或其子目录，右击选择 New→Idl File，如图 2-57 所示。

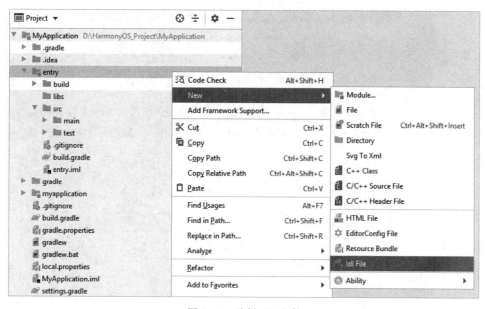

图 2-57　选择 .idl 文件

（2）创建 IDL File。用户既可以直接输入 IDL 接口名称，也可以通过包名格式定义.idl 接口名称，两种方式的差异仅在于.idl 文件的目录结构。

按名称创建 IDL File 时，输入接口名称，直接单击 OK 按钮，如图 2-58 所示。

图 2-58　创建 IDL File

DevEco Studio 在 module 的 src→main 路径下生成 IDL 文件夹，并按照对应模块的包名生成同样的目录结构及 IDL 文件，如图 2-59 所示。

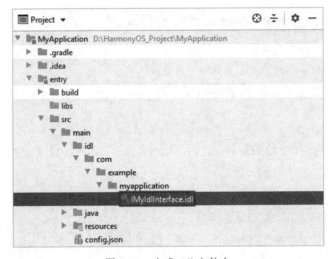

图 2-59　生成.idl 文件夹

按包名创建，自定义.idl 文件存储路径和接口名称。创建 IDL File 时，按照包名创建 IDL 文件。包名利用"."作为分隔符，输入 com. huawei. test. MyIdlInterface，如图 2-60 所示。

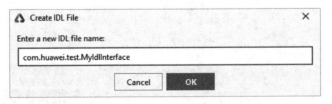

图 2-60　输入名称

DevEco Studio 在 module 的 src→main 路径下生成.idl 文件夹,并按照输入的包名生成相应目录结构及 IDL 文件,可以在此路径继续新增 IDL 文件,如图 2-61 所示。

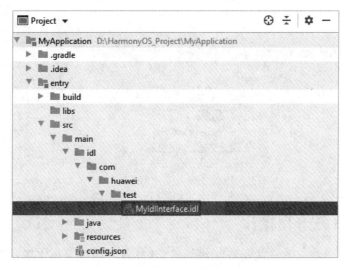

图 2-61　目录结构

(3) 开发者可以使用 Java 或 C++编程语言构建.idl 文件。

(4) 单击工程右栏的 Gradle,在 Tasks→ohos:debug 或 ohos:release 中选择 compileDebugIdl 或 compileReleaseIdl,对模块下的 IDL 文件进行编译,如图 2-62 所示。

图 2-62　编译文件

(5) 编译完成后,在 build→generated→source→idl→{Package Name}目录下,生成对应的接口类、桩类和代理类,如图 2-63 所示。

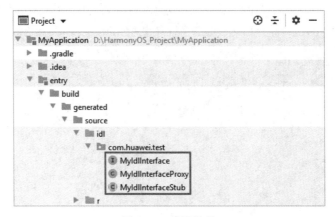

图 2-63　编译结果

2.5.6　服务卡片操作

服务卡片是 FA 的一种主要信息呈现形式,开发者可以在卡片中展示用户最关心的数据,并可以通过单击卡片内容直接打开 FA。例如,天气类 FA,可以在服务卡片中展示当前的基本天气信息,单击卡片启动天气 FA,展示详细的天气数据。同时,服务卡片支持不同的规格尺寸,根据展示的不同内容和布局效果,选用不同的卡片尺寸,支持的尺寸包括：1×2、2×2、2×4 和 4×4。

服务卡片提供了多种类型的模板,开发者可以根据需要展示的信息类型灵活选择模板,快速构建服务卡片,如表 2-6 所示。

表 2-6　模板名称及功能描述

模 板 名 称	功 能 描 述
Grid Pattern（宫格卡片模板）	宫格卡片模板在大尺寸的卡片上特征较为明显,能够有规律进行布局排列。例如展示多排应用图标,每个热区可独立单击或展示影视海报等信息,以凸显图片为主,描述文本为辅
Image With Infomation（图文卡片模板）	图文卡片模板主要在于展现图片和一定数量文本的搭配,在这种布局下,图片和文本属于同等重要信息。在不同尺寸下,图片大小和文本数量会发生一定变化,用于凸显关键信息
Immersive Pattern（沉浸布局卡片模板）	图片内容更能够吸引用户的展现形式,因此沉浸式的布局能够拥有更好的代入感和展现形式。相比较图文和宫格类,这种布局在造型上的制约会更小,设计形式上的发挥空间更大,但在不同设备下的适配需要注意展示效果
List Pattern（列表卡片模板）	列表卡片模板是展示信息时的常用界面组件,通常会在列表的左侧或右侧带有图片或点缀元素。这类布局的优势在于可以集中展示较多信息量,并遵循有序的排列。常用于新闻类、搜索类应用,方便用户获取关键的文本信息

续表

模 板 名 称	功 能 描 述
Circular Data(环形数据模板)	环形数据模板主要用于展示自定义内容数据,卡片主体由环形数据图和文本描述组成,用于凸显关键数据的所占比例
Immersive Data(沉浸式数据模板)	此类型卡片在沉浸式图片上呈现数据信息,可以使用不同的图标搭配信息进行呈现,强调使用场景与数据之间的关系,开发者可以发挥图文搭配的优势,创造出独特风格的卡片样式
Immersive Information(沉浸式图文模板)	沉浸式图文模板的装饰性较强,能够较好提升卡片品质感并起到装饰桌面的作用,合理的布局信息与背景图片之间的空间比例,可以提升用户的个性化使用体验
Multiple Contacts(多个联系人模板)	多个联系人信息融合在一张卡片中,用户能够快捷地查找到最近通话的联系人。也可以通过赋予卡片编辑能力,为用户提供动态可自定义的联系人卡片
Multiple Functions(多功能模板)	开发者可以定义此卡片不同热区位置的单击事件,可以执行某一指令或者不同功能界面的跳转。权衡多个功能之间重要程度,将较大的空间位置留给主要的信息,搭配图片使用,使卡片内容看起来更加丰富
Music Player(音乐播放器卡片模板)	音乐播放器卡片模板主要用于在桌面展示一个音乐播放的控制界面,通过单击卡片上的对应功能按钮,能够实现对音乐播放的控制
Schedule(行程卡片模板)	行程卡片模板主要用于在卡片上展示行程关键信息,并带有功能图标,可通过单击功能图标查看详细行程信息
Shortcuts(捷径卡片模板)	捷径卡片模板主要用于在桌面展示多个快捷功能图标,在这种布局下,每个热区独立可单击,可快速进入相关功能。但需要提供对用户有价值、有服务场景的功能,不要滥用卡片的入口位置
Social Call(通话卡片模板)	通话卡片模板主要用于在桌面显示自定义的联系人图片和通话按钮,在这种场景下,可以直接单击卡片上的通话按钮进行快速呼叫
Standard Image(标准图文模板)	标准图文模板使用场景较广泛,图片和文字信息类型展示基本都可以使用此模板,但仍然是呈现图片为主,例如展示二维码信息、乘车路线图、卡片信息预览等
Standard List(标准列表模板)	此卡片的优势在于强调标题信息,且有序排列,可以明确呈现主副信息内容。列表类型的卡片需要有计划地使用,避免整张卡片全是文字信息
Timer Progress(标准时间进度模板)	此卡片主要突出时间数据,配合标题及正文对数据信息进行解释。可以使用不同的色彩强调信息的重要性,突出核心的内容

服务卡片使用约束如下:只有 Phone、Tablet 和 Wearable 设备的 FA 支持服务卡片,每个 FA 最多可以配置 16 张服务卡片,JS 卡片不支持调试。

1. 创建服务卡片

DevEco Studio 提供服务卡片的一键创建功能,可以快速创建和生成服务卡片模板。

对于创建新工程，可以在工程向导中勾选 Show in Service Center，该参数表示是否在服务中心显示。如果 Project Type 为 Service，则会同步创建一个 2×2 的服务卡片模板，同时还会创建入口卡片。如果 Project Type 为 Application，则只会创建一个 2×2 的服务卡片模板，如图 2-64 所示。

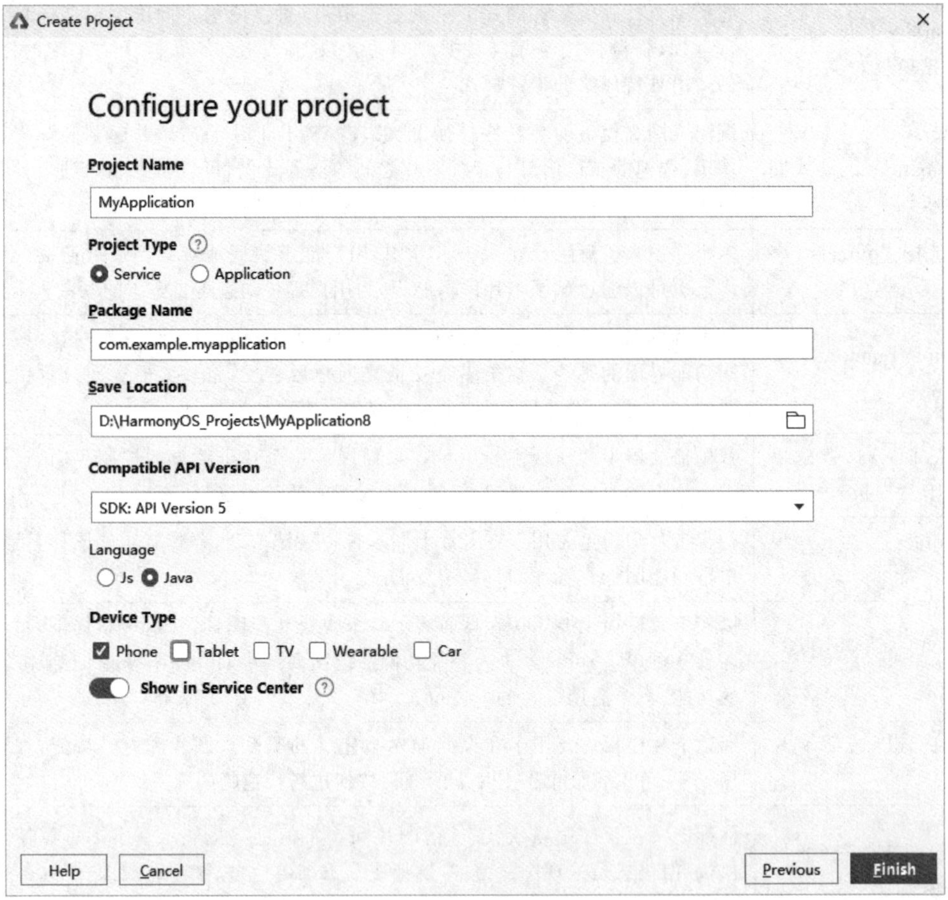

图 2-64　创建服务卡片

卡片创建完成后，会在工程目录下生成 EntryCard 目录，如图 2-65 所示。

在该目录下，每个拥有 EntryCard 的模块，都会生成一个和模块名相同的文件夹，同时还会默认生成一张 2×2 的快照型 EntryCard 图片（.png 格式）。

开发者可以将其替换为提前设计好的 2×2 快照图：将新的快照图复制到目录下，删除默认图片，新图片命名遵循格式"卡片名称-2 * 2.png"。

在已有工程中添加新模块，也可以添加服务卡片和 EntryCard，只需在创建模块时，勾选 Show in Service Center 即可。创建后的服务卡片和 EntryCard 同创建新工程生成的一致。

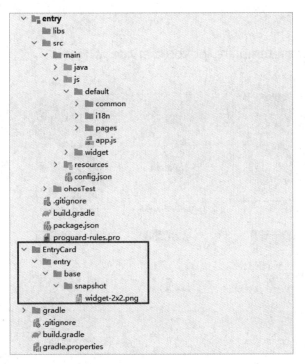

图 2-65 生成 EntryCard 目录

在已有工程中添加 EntryCard,只能通过手工方式,按照 EntryCard 目录创建对应的文件夹和图片。

2. 在已有工程中新添加服务卡片

在已有工程中,新添加服务卡片步骤如下。

(1) 打开一个工程,创建服务卡片模板,创建方法如下。

选择模块(如 entry)下的任意文件,单击菜单栏 File→New→Service Widget,创建服务卡片。

选择模块(如 entry)下的任意文件,右击选择 New→Service Widget,创建服务卡片。

(2) 在 Choose a template for your service widget 界面中,选择需要创建的卡片模板,单击 Next 按钮,如图 2-66 所示。

(3) 在 Configure Your Service Widget 界面中,配置卡片的基本信息如下。

Service Widget Name:卡片名称,在同一个 FA 中不能重复,且只能包含数字、字母和下画线。

Description:卡片的描述信息。

Select Ability/New Ability:选择一个挂靠服务卡片的 Page Ability,或者创建一个新的 Page Ability。

Type:卡片的开发语言类型。

JS Component Name:Type 选择 JS 时需要设置卡片的 JS Component 名称。

图 2-66　创建卡片模板

Support Dimensions：选择卡片的规格，同时还可以查看、预览卡片的效果图。部分卡片支持同时设置多种规格，如图 2-67 所示。

(4) 单击 Finish 按钮完成卡片的创建。创建完成后，工具会自动创建出服务卡片的布局文件，并在 config.json 文件中写入服务卡片的属性字段，如图 2-68 所示。

(5) 创建完成后进行服务卡片的开发。

3. 预览服务卡片

在开发服务卡片过程中，支持对卡片进行实时预览。服务卡片通过 XML 或 JS 文件进行布局设计，可以对布局 XML/JS 文件进行实时预览，只要在 XML/JS 布局文件中保存了修改的源代码，在预览器中就可以实时查看布局效果。在 Phone 和 Tablet 服务卡片的预览效果中，每个尺寸的服务卡片提供 3 种预览效果，分别为极窄(Minimum)、标准(Default)和极宽(Maximum)，开发者应确保 3 种尺寸的显示效果均正常，以便适应不同屏幕尺寸的设备，如图 2-69 所示。

第2章 应用开发基础 77

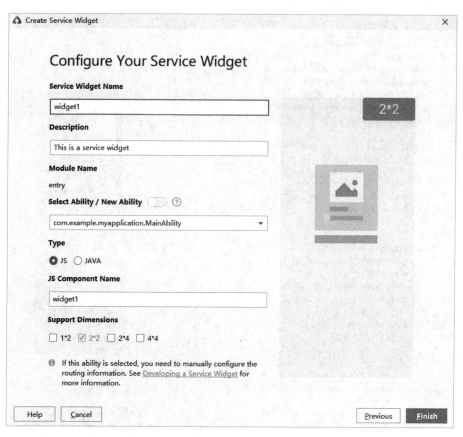

图 2-67 卡片规格

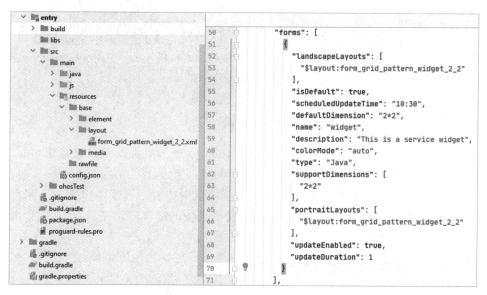

图 2-68 添加卡片属性

图 2-69 预览服务卡片

2.5.7 使用预览器查看应用效果

在 HarmonyOS 应用开发过程中，DevEco Studio 为开发者提供了预览器的功能，可以查看应用的 UI 效果，支持 Java 应用和 JS 应用的预览。预览器支持布局代码的实时预览，只需要将开发的源代码进行保存，就可以通过预览器实时查看运行效果，方便开发者随时调整代码。需要注意的是，由于 Windows 系统和真机设备的字体库存在差异，可能会出现预览器界面中的字体与真机运行效果的字体存在差异。

API 5/API 6 的预览功能如表 2-7 所示。为了获得更好的使用体验，建议先将 DevEco Studio 升级至最新版本，然后在 SDK Manager 中检测并更新 SDK 至最新版本。

表 2-7　API 5/API 6 预览功能

JS(Windows)	JS(macOS)	Java(Windows)	Java(macOS 版)
实时预览：支持	实时预览：支持	实时预览：支持	实时预览：支持
动态预览：支持	动态预览：支持	动态预览：仅支持 Java 代码布局，不支持 XML 布局	动态预览：仅支持 Java 代码布局，不支持 XML 布局
Profile Manager：支持	Profile Manager：支持	Profile Manager：支持	Profile Manager：支持
多端设备预览：支持，支持的设备包括 Wearable、TV、Phone、Tablet 和 Lite Wearable	多端设备预览：支持，支持的设备包括 Wearable、TV、Phone、Tablet 和 Lite Wearable	多端设备预览：支持，支持的设备包括 Wearable、TV、Phone、Car 和 Tablet	多端设备预览：支持，支持的设备包括 Wearable、TV、Phone、Car 和 Tablet
双向预览：支持，支持的设备包括 Wearable、TV、Phone 和 Tablet；仅支持单设备，不支持多端设备场景	双向预览：支持，支持的设备包括 Wearable、TV、Phone 和 Tablet；仅支持单设备，不支持多端设备场景	双向预览：支持，支持的设备包括 Wearable、TV、Phone、Tablet 和 Car；仅支持单设备，不支持多端设备场景；仅支持 XML 布局预览场景，不支持 Java 代码布局场景	双向预览：支持，支持的设备包括 Wearable、TV、Phone、Tablet 和 Car；仅支持单设备，不支持多端设备场景；仅支持 XML 布局预览场景，不支持 Java 代码布局场景

1. 查看 JS 应用预览效果

在使用 JS 预览器前，检查如下环境信息：确保 HarmonyOS Legacy SDK→SDK Tools 中，已下载 Previewer 资源。如果已下载但存在新版本的情况，建议升级到最新版本。HarmonyOS Legacy SDK→SDK Platform 中的 JS SDK 需更新至最新版本，使用 JS 预览器方法如下。

（1）创建或打开一个 JS 应用工程。

（2）在创建的工程目录下，打开任意一个页面下的 hml\\css\\js 文件。

（3）可以通过如下任意一种方式打开预览器开关，获得显示效果。

单击菜单栏 View→Tool Windows→Previewer 或使用快捷键 Alt+3(Mac 为 Option+3)，打开预览器。在编辑窗口右上角的侧边工具栏，单击 Previewer 按钮，打开预览器。

可以实现实时预览，只要在布局文件中保存修改的源代码，在预览器中就可以实时查看布局效果。也可以实现动态预览，在预览器界面，可以操作应用的交互动作，例如单击事件、跳转、滑动等，与应用运行在真机设备上的交互体验一致。

2. 查看 Java 应用预览效果

在使用 Java 预览器前，检查如下环境信息。

需要确保 HarmonyOS Legacy SDK→SDK Tools 中，已下载 Previewer 资源，如果已下载但存在新版本的情况，需要升级到最新版本。HarmonyOS Legacy SDK→SDK Platform 中的 Java SDK 需要更新至最新版本，使用 Java 预览器的方法如下。

（1）创建或打开一个 Java 应用工程。

(2) 在创建的工程目录下,根据布局方式,打开布局文件如下。

JavaUI 布局:打开一个 AbilitySlice.java 或 Ability.java 文件。

Windows 系统:在 File→Settings→DevEco Labs→Previewer 下勾选 Enable Java Previewer 开启。

macOS 系统:在 DevEco Studio→Preferences→DevEco Labs→Previewer 下勾选 Enable Java Previewer 开启。

XML 布局:打开 resources→base→layout 目录下的 XML 布局文件。

(3) 可以通过如下任意一种方式打开预览器开关,获取显示效果。

单击菜单栏 View→Tool Windows→Previewer 或使用快捷键 Alt+3(Mac 为 Option+3),打开预览器。在编辑窗口右上角的侧边工具栏,单击 Previewer 按钮,打开预览器。

Java 预览器支持 Phone、Tablet、Car、TV 和 Wearable 设备的 Java 应用布局预览。Java 应用的布局支持 Java 代码布局和 XML 布局两种方式,其中 Java 代码布局(AbilitySlice.java 或 Ability.java 文件)支持实时预览界面布局效果,同时还可以动态预览应用的交互效果,例如单击、跳转、滑动等互动式操作;XML 布局文件可以实时预览、修改和保存 XML 代码后,预览器会实时展示应用的布局效果。

3. Profile Manager

由于真机设备有丰富的型号,例如 Phone 设备包括 Mate30、Mate40、P40、P50 等,不同设备型号的屏幕分辨率可能不同。因此,在原子化服务或 HarmonyOS 应用开发过程中,由于设备类型繁多,可能无法查看在不同设备上的界面显示效果。对此,DevEco Studio 的预览器提供了 Profile Manager 功能,支持开发者自定义预览设备 Profile(包含分辨率和语言),从而可以通过定义不同的预览设备 Profile,查看 HarmonyOS 应用或原子化服务在不同设备上的预览显示效果。当前支持自定义设备分辨率及系统语言,如果是 Lite Wearable 设备类型,还支持自定义屏幕形状。定义设备后,可以在 Previewer 中,单击设备型号,切换预览设备。

下面以自定义一款 Phone 设备为例,介绍 Profile Manager 的使用方法。

(1) 在预览器界面,打开 Profile Manager。

(2) 在 Profile Manager 界面,单击"+"按钮,添加设备。

(3) 在 Create Profile 界面,填写新增设备的信息,例如 Profile ID(设备型号)、Device type(设备类型)、Resolution(分辨率)和 Language(语言)等。其中,Device type 只能选择 config.json 中 deviceType 字段已定义的设备,如图 2-70 所示。

(4) 设备信息填写完成后,单击 OK 按钮完成创建。

4. 查看多端设备预览效果

DevEco Studio 支持 HarmonyOS 分布式应用开发,同一个应用可以运行在多个设备上。在 HarmonyOS 分布式应用的开发阶段,因不同设备的屏幕分辨率、形状、大小不同,开发者需要在不同的设备上查看应用的 UI 布局和交互效果,此时便可以使用多端设备预览器功能,在开发过程中随时查看不同设备上的运行效果,多端设备预览最多同时支持 4 个设

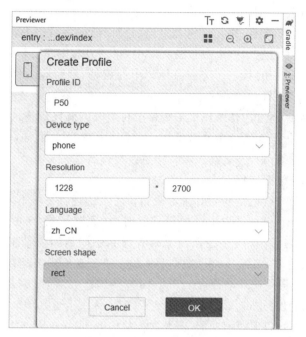

图 2-70　填写新增设备信息

备的预览。

前面介绍了 DevEco Studio 支持 JS 应用和 Java 应用的预览器功能,多端设备预览器支持 JS 应用与 Java 应用在不同设备上的同时预览。如果两个设备支持的编码语言不同,就不能使用多端设备预览功能,例如 Java 语言 Phone 的应用是不支持在 LiteWearable 上运行的,因为 LiteWearable 不支持 Java 语言。

Java 应用的多端设备预览器使用方法如下。

(1) 在工程目录中,打开任意一个 HML 页面(Java 打开 AbilitySlice.java 或 Ability.java 文件)。

(2) 可以通过如下任意一种方式打开预览器开关,获得显示效果。

单击菜单栏 View→Tool Windows→Previewer,打开预览器。在编辑窗口右上角的侧边工具栏,单击 Previewer 按钮,打开预览器。

(3) 在 Previewer 窗口中,打开 Profile Manager 中的 Multi-profile preview 开关,同时查看多设备上的应用运行效果。

多端设备预览不支持动画预览,如果需要查看动画在设备上的预览效果,需关闭 Multi-device preview 功能后,在单设备预览界面进行查看。

5. 双向预览

DevEco Studio 预览器支持 Java 的 XML 布局和 JS 的双向预览功能。使用双向预览功能时,需要在预览器界面单击图标打开此功能。开启后,支持代码编辑器、UI 和 Component

Tree 控件树三者之间的联动。

选中预览器 UI 中的控件,则控件树上对应的控件组件将被选中,同时代码编辑器布局文件中对应的代码块高亮显示。选中布局文件中的代码块,则在 UI 中会高亮显示,控件树上的节点也会呈现被选中的状态。选中控件树中的控件,则对应的代码块和 UI 也会高亮显示。

在预览界面还可以通过控件的属性面板修改可修改的属性或样式,在预览界面修改后,预览器会自动同步到代码编辑器中修改源码,并实时刷新 UI。同样,如果在代码编辑器中修改源码,也会实时刷新 UI,并更新控件树信息及控件属性。

6. PreviewMock 数据模拟

在预览场景中,由于代码的运行环境与真机设备上的运行环境不同,调用部分接口时无法获取到有效的返回值,例如获取电池电量信息等,在预览场景下 getVoltage() 返回的是一个固定的值 0,这样开发者就无法在预览时查看到不同返回值带来的界面变化。因此,DevEco Studio 提供了 PreviewMock 数据模拟功能,在不改变业务运行逻辑的同时,开发者可以模拟 API 或者业务代码中各种 method(不包括构造方法)的返回值和对象中的 Field(不包括 final 字段)的值。

使用 PreviewMock 数据模拟功能,需要在模块的 build.gradle 中添加其依赖,然后重新同步工程。

```
dependencies {
    ...
    implementation group: 'com.huawei.deveco', name: 'previewer-mock-core', version: '1.0.0.1'
}
```

1) Method 的 Mock

Method 的 Mock 步骤如下。

(1) 在源码目录下新建一个 class,该 class 继承自 com.huawei.asm.core.PreviewerMock。

```
public class MockClassB extends PreviewMock{
}
```

(2) 在 class 中添加 com.huawei.asm.core.annotation.PreviewerMockMethod 注解,然后定义一个和原方法同名(支持 public、private、protected、static 和 final)的方法。

```
public class MockClassB extends PreviewMock{
    //对方法返回值的 Mock,例如对一个 BatteryInfo 类的 getVoltage()方法的 Mock
    @PreviewerMockMethod
    public int getVoltage(BatteryInfo batteryInfo) {    //第一个参数为原方法的对象,后面的参
                                                        //数为原方法本身参数
        return 30;
    }
    //对方法入参的 Mock,例如对 MainAbilitySlice 的 onStart 方法的 Mock
    @PreviewerMockMethod
```

```
public void onStart(MainAbilitySlice mainAbilitySlice, Intent intent) {
                        //第一个参数为原方法的对象,后面的参数为原方法本身参数
    intent.setBundle("mock bundle Name");
    mainAbilitySlice.onStart(intent);
}
```

（3）在原方法中添加 Hilog 日志,方便预览时,在 PreviewerLog 中打印获取返回值,从而验证 Mock 是否生效。

```
//在获取电量信息的方法中添加 Hilog 日志
HiLog.debug(hiLoglabel, batteryInfo.getVoltage() + "");
//在 onStar 方法中添加 Hilog 日志
HiLog.debug(hiLoglabel, intent.getBundle() + "");
```

2）对 Field 值的 Mock

对 Field 值的 Mock 分为 private/public/protected 非静态字段和 static 字段两种。

（1）对 private/public/protected 非静态字段的 Mock。在原代码调用 Field 之前,添加 Mock 代码,格式为 public static < T > void mockField(Object ref, String fieldName, T fieldValue)。其中,第一个参数为要 mock 的对象,第二个参数为要 mock 的字段名称,第三个参数为要 mock 的值。然后在 HiLog 中打印日志,验证 Mock 是否生效。

```
PreviewerFieldMock.mockField(sensorData, "accuracy", 20);
HiLog.debug(hiLoglabel, sensorData.accuracy + "");
```

（2）对 static 静态字段的 Mock。在原代码调用 Field 之前,添加 Mock 代码,格式为 public static< T > void mockStaticField(Class <?> obj, String fieldName, T fieldValue)。第一个参数为要 Mock 的 Class,第二个参数为要 Mock 的字段名称,第三个参数为要 Mock 的值。然后在 HiLog 中打印日志,验证 Mock 是否生效。

```
PreviewerFieldMock.mockStaticField(Demo.class, "staticName", "mock static name");
HiLog.debug(hiLoglabel, "mock Demo Static : " + Demo.staticName);
```

2.5.8 将 SVG 文件转换为 XML 文件

SVG(Scalable Vector Graphics)可缩放矢量图形,是一种图像文件格式。目前由于 HarmonyOS 图形渲染引擎不支持 SVG 格式,开发者需要将 SVG 格式的图片转换为 XML 格式的文件,然后在布局文件中引用转换后的 XML 文件,这样,在模拟器/预览器或者设备上运行应用时,能够正常渲染该图像文件,转换方法如下。

（1）选中应用模块,右击选择 New→Svg To Xml。

（2）选择需要转换的 SVG 文件并命名,单击 OK 按钮开始转换。

（3）转换成功后,可以在 resources→base→graphic 文件下找到转换后的 XML 文件,在布局文件中引用 XML 文件名即可完成对图标文件的引用。

第 3 章 Ability 框架开发——基于 Java

本章将基于 Java 对 Ability、公共事件与通知开发、后台任务调度和管控、线程管理开发、线程间通信、剪贴板开发进行介绍。受限于本书的篇幅，基于 JS 的 Ability 开发随书附赠。

3.1 开发概述

HarmonyOS 应用开发主要包括通用开发和原子化服务开发。其中，通用开发包括开发 Ability、开发 UI 和开发业务功能。

1. 开发 Ability

进行 HarmonyOS 应用开发要了解 Ability 如何使用。Ability 是 HarmonyOS 应用程序的重要组成部分，分为 FA 和 PA 两种类型。

FA 支持 Page Ability：Page 模板是 FA 唯一支持的模板，用于提供与用户交互的能力。

PA 支持 Service Ability 和 Data Ability：Service 模板用于提供后台运行任务的能力；Data 模板用于对外部提供统一的数据访问抽象。

每种类型为开发者提供了不同的模板，以便实现不同的业务功能。一个 Page 实例可以包含一组相关页面，每个页面用一个 AbilitySlice 实例表示。

2. 开发 UI

FA 需要提供 UI 用于与用户进行交互，HarmonyOS 提供 Java UI 和 JS UI 两种 UI 框架：Java UI 提供细粒度的 UI 编程接口，使应用开发更加灵活；JS UI 提供相对高层的 UI 描述，使应用开发更加简单。针对轻量级智能穿戴（Lite Wearable），现阶段只使用 JS 语言进行应用开发，示例工程参考地址为：https://developer.harmonyos.com/cn/docs/documentation/doc-guides/lite-wearable-experience-0000000000622606。

3. 开发业务功能

媒体：视频、音频、图像、相机等功能的开发。

安全：权限、生物特征识别等功能的开发。

AI：图像超分、语音识别等功能的开发。

网络连接：NFC、蓝牙、WLAN 等功能的开发。
设备管理：传感器、控制类小器件、位置等功能的开发。
数据管理：数据库、分布式数据/文件服务、数据搜索等功能的开发。
线程：线程管理、线程间通信等功能的开发。
IDL：声明系统服务和 Ability 对外提供的服务接口，并生成相关代码。

4. 原子化服务开发

HarmonyOS 除支持传统方式需要安装应用外，还支持提供特定功能的免安装应用（原子化服务），供用户在合适的场景和设备上便捷使用。

原子化服务相对于传统方式需要安装的应用更加轻量，同时提供丰富的入口、更精准的分发。

3.2 Ability 介绍

Ability 是应用所具备能力的抽象，也是应用程序的重要组成部分。一个应用可以具备多种能力（可以包含多个 Ability），HarmonyOS 支持应用以 Ability 为单位进行部署。

在配置文件（config.json）中注册 Ability 时，可以通过配置 Ability 元素中的 type 属性指定 Ability 模板类型，相关代码如下。

```
{
    "module": {
        ...
        "abilities": [
            {
                ...
                "type": "page"
                ...
            }
        ]
        ...
    }
    ...
}
```

其中，type 的取值可以为 page、service 或 data，分别代表 Page、Service 和 Data 模板。为便于表述，后续将基于 Page 模板、Service 模板、Data 模板实现的 Ability 分别简称为 Page、Service 和 Data。

3.2.1 Page Ability

本部分包括 Page 与 AbilitySlice、AbilitySlice 路由配置、Page Ability 生命周期、AbilitySlice 间导航和跨设备迁移。

1. Page 与 AbilitySlice

Page 模板(以下简称 Page)是 FA 唯一支持的模板,用于提供与用户交互的能力。一个 Page 可以由一个或多个 AbilitySlice 构成,AbilitySlice 是指应用的单个页面及其控制逻辑的总和。

当一个 Page 由多个 AbilitySlice 共同构成时,这些 AbilitySlice 页面提供的业务能力应具有高度相关性。例如,新闻浏览功能可以通过一个 Page 实现,其中包含了两个 AbilitySlice:一个用于展示新闻列表,另一个用于展示新闻详情。Page 与 AbilitySlice 的关系如图 3-1 所示。

图 3-1　Page 与 AbilitySlice 的关系

相比桌面场景,移动场景下应用之间的交互更为频繁。通常,单个应用专注于某个方面的能力开发,当需要其他能力辅助时,会调用其他应用提供的能力。例如,外卖应用提供了联系商家的业务功能入口,当用户在使用该功能时,会跳转到通话应用的拨号页面。与此类似,HarmonyOS 支持不同 Page 之间的跳转,并可以指定跳转到目标 Page 中某个具体的 AbilitySlice。

2. AbilitySlice 路由配置

虽然一个 Page 可以包含多个 AbilitySlice,但是 Page 进入前台时界面默认只展示一个 AbilitySlice。默认展示的 AbilitySlice 是通过 setMainRoute()方法指定的。如果需要更改默认展示的 AbilitySlice,可以通过 addActionRoute()方法配置一条路由规则。此时,当其他 Page 实例期望导航到此 AbilitySlice 时,可以在 Intent 中指定 Action。

setMainRoute()方法与 addActionRoute()方法的代码如下。

```java
public class MyAbility extends Ability {
    @Override
    public void onStart(Intent intent) {
        super.onStart(intent);
        //设置主路由
        setMainRoute(MainSlice.class.getName());
        //设置动作路由
        addActionRoute("action.pay", PaySlice.class.getName());
        addActionRoute("action.scan", ScanSlice.class.getName());
    }
}
```

addActionRoute()方法中使用的动作命名,需要在应用配置文件(config.json)中注册。

```
{
    "module": {
        "abilities": [
            {
                "skills":[
                    {
                        "actions":[
                            "action.pay",
                            "action.scan"
                        ]
                    }
                ]
                ...
            }
        ]
        ...
    }
    ...
}
```

3. Page Ability 生命周期

系统管理或用户操作等行为均会引起 Page 实例在其生命周期的不同状态之间进行转换。Ability 类提供的回调机制能够让 Page 及时感知外界变化,从而正确地应对状态变化(释放资源),有助于提升应用的性能。

1) Page 生命周期回调

Page 生命周期的不同状态转换及其对应的回调如图 3-2 所示。

(1) onStart()。当系统首次创建 Page 实例时,触发该回调。对于一个 Page 实例,回调在其生命周期过程中仅触发一次,Page 在逻辑后进入 INACTIVE 状态。开发者必须重写此方法,并配置默认展示的 AbilitySlice。

```
@Override
public void onStart(Intent intent) {
    super.onStart(intent);
    super.setMainRoute(FooSlice.class.getName());
}
```

(2) onActive()。Page 会在进入 INACTIVE 状态后到前台,然后系统调用此回调。Page 在此之后进入 ACTIVE 状态,它是应用与用户交互的状态。Page 将保持在此状态,除非某类事件发生导致 Page 失去焦点,例如用户单击返回按钮或导航到其他 Page。当此类事件发生时,会触发 Page 回到 INACTIVE 状态,系统将调用 onInactive()回调。此后 Page 可能重新回到 ACTIVE 状态,系统将再次调用 onActive()回调。因此,开发者通常需要成对实现 onActive()和 onInactive(),并在 onActive()中获取在 onInactive()中被释放的

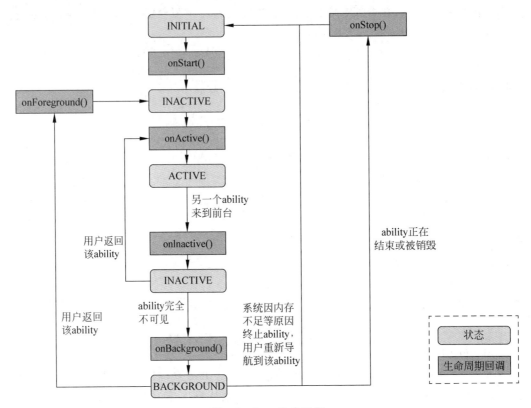

图 3-2 Page 生命周期

资源。

（3）onInactive()。当 Page 失去焦点时，系统将调用此回调，此后 Page 进入 INACTIVE 状态。开发者可以在此回调中实现 Page 失去焦点时应表现的恰当行为。

（4）onBackground()。如果 Page 不再对用户可见，系统将调用此回调通知用户进行相应的资源释放，此后 Page 进入 BACKGROUND 状态。开发者应在此回调中释放 Page 不可见时无用的资源，或在此回调中执行较为耗时的状态保存操作。

（5）onForeground()。处于 BACKGROUND 状态的 Page 仍然驻留在内存中，当重新回到前台时（用户重新导航到此 Page），系统先调用 onForeground() 回调通知开发者，而后 Page 的生命周期状态回到 INACTIVE 状态。开发者应当在此回调中重新申请 onBackground() 中释放的资源，最后 Page 的生命周期回到 ACTIVE 状态，系统通过 onActive() 回调通知开发者用户。

（6）onStop()。系统要销毁 Page 时，会触发此回调函数，通知用户进行系统资源的释放。销毁 Page 的可能原因包括以下几方面：用户通过系统管理能力关闭指定 Page，例如使用任务管理器关闭 Page。用户行为触发 Page 的 terminateAbility() 方法调用，例如使用应用的退出功能。配置变更导致系统暂时销毁 Page 并重建。系统出于资源管理目的，自动触发对处于 BACKGROUND 状态 Page 的销毁。

2）AbilitySlice 生命周期

AbilitySlice 作为 Page 的组成单元，其生命周期是依托于所属 Page 生命周期的。AbilitySlice 和 Page 具有相同的生命周期状态和同名的回调，当 Page 生命周期发生变化时，它的 AbilitySlice 也会发生相同的生命周期变化。此外，AbilitySlice 还具有独立于 Page 的生命周期变化，发生在同一 Page 中的 AbilitySlice 之间导航时，此时 Page 的生命周期状态不会改变。

AbilitySlice 生命周期回调与 Page 的相应回调类似，因此不再赘述。由于 AbilitySlice 承载具体的页面，开发者必须重写 AbilitySlice 的 onStart() 回调，并在此方法中通过 setUIContent() 方法设置页面，相关代码如下。

```
@Override
protected void onStart(Intent intent) {
    super.onStart(intent);
    setUIContent(ResourceTable.Layout_main_layout);
}
```

AbilitySlice 实例创建和管理通常由应用负责，系统仅在特定情况下创建 AbilitySlice 实例。例如，通过导航启动某个 AbilitySlice 时，由系统负责实例化，但在同一个 Page 中不同的 AbilitySlice 间导航时则由应用负责实例化。

3）Page 与 AbilitySlice 生命周期关联

当 AbilitySlice 处于前台且具有焦点时，其生命周期状态随着所属 Page 的生命周期状态的变化而变化。一个 Page 拥有多个 AbilitySlice 时，例如 MyAbility 下有 FooAbilitySlice 和 BarAbilitySlice，当前 FooAbilitySlice 处于前台且获得焦点，并将导航到 BarAbilitySlice，在此期间的生命周期状态变化顺序如下。

FooAbilitySlice 从 ACTIVE 状态变为 INACTIVE 状态。BarAbilitySlice 则从 INITIAL 状态首先变为 INACTIVE 状态，然后变为 ACTIVE 状态（假定此前 BarAbilitySlice 未曾启动）。FooAbilitySlice 从 INACTIVE 状态变为 BACKGROUND 状态。

对应两个 Slice 的生命周期方法回调顺序如下：FooAbilitySlice.onInactive()→BarAbilitySlice.onStart()→BarAbilitySlice.onActive()→FooAbilitySlice.onBackground()。

在整个流程中，MyAbility 始终处于 ACTIVE 状态。但是，当 Page 被系统销毁时，其所有已实例化的 AbilitySlice 将联动销毁，而不仅是处于前台的 AbilitySlice。

4．AbilitySlice 间导航

本部分介绍同一 Page 内导航和不同 Page 间导航。

1）同一 Page 内导航

当发起导航的 AbilitySlice 和导航目标的 AbilitySlice 处于同一个 Page 时，通过 present() 方法实现导航。单击按钮导航到其他 AbilitySlice 的相关代码如下。

```
@Override
```

```
protected void onStart(Intent intent) {
    ...
    Button button = ...;
    button.setClickedListener(listener -> present(new TargetSlice(), new Intent()));
    ...
}
```

如果希望在用户从导航目标 AbilitySlice 返回时能够获得其返回结果，则应当使用 presentForResult() 实现导航。用户从导航目标 AbilitySlice 返回时，系统将回调 onResult() 接收和处理返回结果，开发者需要重写该方法。返回结果由导航目标 AbilitySlice 在其生命周期内通过 setResult() 进行设置，相关代码如下。

```
int requestCode = positiveInteger; //任何正整数
@Override
protected void onStart(Intent intent) {
    ...
    Button button = ...;
    button.setClickedListener(
        listener -> presentForResult(new TargetSlice(), new Intent(), positiveInteger));
    ...
}
@Override
protected void onResult(int requestCode, Intent resultIntent) {
    if (requestCode == positiveInteger) {
        //在此处理 resultIntent
    }
}
```

系统为每个 Page 维护了一个 AbilitySlice 实例的栈，每个进入前台的 AbilitySlice 实例均会入栈。当开发者在调用 present() 或 presentForResult() 时，若指定的 AbilitySlice 实例已经在栈中存在，则栈中位于此实例之上的 AbilitySlice 均会出栈并终止其生命周期。前面的示例代码中，导航时指定的 AbilitySlice 实例均是新建的，即便重复执行此代码（此时作为导航目标的这些实例是同一个类），也不会导致任何 AbilitySlice 出栈。

2) 不同 Page 间导航

AbilitySlice 作为 Page 的内部单元，以 Action 的形式对外暴露，因此通过配置 Intent 的 Action 导航到目标 AbilitySlice。Page 间的导航使用 startAbility() 或 startAbilityForResult() 方法，获得返回结果的回调为 onAbilityResult()。在 Ability 中调用 setResult() 可以设置返回结果。

5．跨设备迁移

跨设备迁移（以下简称"迁移"）支持将 Page 在同一用户的不同设备间迁移，以便支持用户无缝切换的需求，开发步骤如下。

1) 实现 IAbilityContinuation 接口

一个应用可能包含多个 Page，仅需要在支持迁移的 Page 中通过以下方法实现 IAbilityContinuation 接口。同时，此 Page 所包含的所有 AbilitySlice 也需要实现此接口。

onStartContinuation()：Page 请求迁移后，系统首先回调此方法，开发者可以在此回调中决策当前是否可以执行迁移，例如弹框让用户确认是否开始迁移。

onSaveData()：如果 onStartContinuation() 返回 true，则系统回调此方法，在此回调中保存的数据必须传递到另外设备上，以便恢复 Page 状态的数据。

onRestoreData()：源侧设备上 Page 完成保存数据后，系统在目标侧设备上回调此方法，开发者在此回调中接收用于恢复 Page 状态的数据。注意，在目标侧设备上的 Page 会重新启动其生命周期，无论其启动模式如何配置，且系统回调此方法的时机在 onStart() 之前。

onCompleteContinuation()：目标侧设备上恢复数据一旦完成，系统就会在源侧设备上回调 Page 的方法，以便通知应用迁移流程已结束。开发者可以在此检查迁移结果是否成功，并在此处理迁移结束后的动作，例如应用可以在迁移完成后终止自身生命周期。

onFailedContinuation()：迁移过程中发生异常，系统会在发起端设备上回调 FA 的此方法，以便通知应用迁移流程发生的异常。并不是所有异常都会回调 FA，仅局限于该接口枚举的异常。开发者可以在此检查异常信息，并在此处理迁移异常发生后的动作，例如可以提醒用户此时发生的异常信息。该接口从 API 版本 6 开始提供，目前为 Beta 版本。

onRemoteTerminated()：如果使用 continueAbilityReversibly()，而不是 continueAbility()，则此后可以在源侧设备上使用 reverseContinueAbility() 进行回迁。这种场景下，相当于同一个 Page（两个实例）同时在两个设备上运行，迁移完成后，如果目标侧设备上 Page 因任何原因终止，则源侧 Page 通过此回调接收终止通知。

2) 请求迁移

实现 IAbilityContinuation 的 Page 可以在其生命周期内，调用 continueAbility() 或 continueAbilityReversibly() 请求迁移。二者区别：通过后者发起的迁移，可以进行回迁。

```
try {
    continueAbility();
} catch (IllegalStateException e) {
    //可以继续进行其他处理
    ...
}
```

以 Page 从设备 A 迁移到设备 B 为例，具体流程如下。

(1) 设备 A 上的 Page 请求迁移。

(2) 系统回调设备 A 上的 Page 及其 AbilitySlice 栈中所有 AbilitySlice 实例的 IAbilityContinuation.onStartContinuation() 方法，以确认当前是否可以立即迁移。

(3) 如果可以立即迁移，则系统回调设备 A 上的 Page 及其 AbilitySlice 栈中所有 AbilitySlice 实例的 IAbilityContinuation.onSaveData() 方法，以便保存迁移后恢复状态的数据。

（4）如果保存数据成功，则系统在设备 B 上启动同一个 Page，并恢复 AbilitySlice 栈，然后回调 IAbilityContinuation.onRestoreData()方法，传递此前保存的数据；此后设备 B 上的 Page 从 onStart()开始其生命周期回调。

（5）系统回调设备 A 上的 Page 及其 AbilitySlice 栈中所有 AbilitySlice 实例的 IAbilityContinuation.onCompleteContinuation()方法，通知数据恢复成功与否。

（6）迁移过程中发生异常，系统回调设备 A 上的 Page 及其 AbilitySlice 栈中所有 AbilitySlice 实例的 IAbilityContinuation.onFailedContinuation()方法，通知迁移过程中发生异常，并不是所有异常都会回调 FA 方法，仅局限于该接口枚举的异常。

3）请求回迁

通过 continueAbilityReversibly()请求迁移并完成后，源侧设备上已迁移的 Page 可以发起回迁，以便使用户活动重新回到此设备。

```
try {
    reverseContinueAbility();
} catch (IllegalStateException e) {
    //可能另一个正在进行中
    ...
}
```

以 Page 从设备 A 迁移到设备 B 后并请求回迁为例，具体流程如下。

（1）设备 A 上的 Page 请求回迁。

（2）系统回调设备 B 上的 Page 及其 AbilitySlice 栈中所有 AbilitySlice 实例的 IAbilityContinuation.onStartContinuation()方法，以确认当前是否可以立即迁移。

（3）如果可以立即迁移，则系统回调设备 B 上的 Page 及其 AbilitySlice 栈中所有 AbilitySlice 实例的 IAbilityContinuation.onSaveData()方法，以便保存回迁后恢复状态的数据。

（4）如果保存数据成功，则系统在设备 A 上的 Page 恢复 AbilitySlice 栈，然后回调 IAbilityContinuation.onRestoreData()方法，传递此前保存的数据。

（5）如果数据恢复成功，则系统终止设备 B 上 Page 的生命周期。

3.2.2 Service Ability

基于 Service 模板的 Ability 主要用于后台运行任务（如执行音乐播放、文件下载等），但不提供用户交互界面。Service 可由其他应用或 Ability 启动，即使用户切换到其他应用，Service 仍将在后台继续运行。

Service 是单实例的。在一个设备上，相同的 Service 只会存在一个实例。如果多个 Ability 共用这个实例，只有当与 Service 绑定的所有 Ability 都退出后，Service 才能退出。由于 Service 在主线程中执行，因此 Service 里面的操作时间过长，开发者必须在 Service 中创建新的线程进行处理，防止造成主线程阻塞，应用程序无响应。

1. 创建 Service

创建 Ability 的子类，实现 Service 相关的生命周期方法。Service 也是一种 Ability，Ability 为 Service 提供了以下生命周期方法，开发者可以重写这些方法，添加其他 Ability 请求与 Service Ability 交互时的处理方法。

（1）onStart()：该方法在创建 Service 时调用，用于 Service 的初始化。在 Service 的整个生命周期只会调用一次，调用时传入的 Intent 为空。

（2）onCommand()：在 Service 创建完成之后调用，该方法在客户端每次启动 Service 时都会调用，开发者可以在该方法中做一些调用统计、初始化类的操作。

（3）onConnect()：在 Ability 和 Service 连接时调用，该方法返回 IRemoteObject 对象，开发者可以在回调函数中生成对应 Service 的 IPC 通信通道，以便 Ability 与 Service 交互。Ability 可以多次连接同一个 Service，系统会缓存 Service 的 IPC 通信对象，只有第一个客户端连接 Service 时，系统才会调用 Service 的 onConnect 方法生成 IRemoteObject 对象，而后系统会将同一个 RemoteObject 对象传递至其他连接同一个 Service 的所有客户端，而无须再次调用 onConnect 方法。

（4）onDisconnect()：在 Ability 与绑定的 Service 断开连接时调用。

（5）onStop()：在 Service 销毁时调用。Service 应通过实现此方法清理任何资源，例如关闭线程、注册的侦听器等，相关代码如下。

```java
public class ServiceAbility extends Ability {
    @Override
    public void onStart(Intent intent) {
        super.onStart(intent);
    }
    @Override
    public void onCommand(Intent intent, boolean restart, int startId) {
        super.onCommand(intent, restart, startId);
    }
    @Override
    public IRemoteObject onConnect(Intent intent) {
        return super.onConnect(intent);
    }
    @Override
    public void onDisconnect(Intent intent) {
        super.onDisconnect(intent);
    }
    @Override
    public void onStop() {
        super.onStop();
    }
}
```

2. 注册 Service

Service 需要在应用配置文件中进行注册，注册类型 Type 需要设置为 Service，相关代码如下。

```
{
    "module": {
        "abilities": [
            {
                "name": ".ServiceAbility",
                "type": "service",
                "visible": true
                ...
            }
        ]
        ...
    }
    ...
}
```

3. 启动 Service

通过 startAbility()启动 Service 及对应的停止方法。

1) 启动 Service

Ability 提供 startAbility()方法启动另外一个 Ability。Service 也是 Ability 的一种,同样可以将 Intent 传递给该方法启动 Service。startAbility()不仅支持启动本地 Service,还支持启动远程 Service。

可以通过构造包含 DeviceId、BundleName 与 AbilityName 的 Operation 对象设置目标 Service 信息,参数含义如下。

DeviceId:表示设备 ID。如果是本地设备,则可以直接留空;如果是远程设备,可以通过 ohos.distributedschedule.interwork.DeviceManager 提供的 getDeviceList 获取设备列表。

BundleName:表示包名称。

AbilityName:表示待启动的 Ability 名称。

启动本地设备 Service 的相关代码如下。

```
Intent intent = new Intent();
Operation operation = new Intent.OperationBuilder()
        .withDeviceId("")
        .withBundleName("com.domainname.hiworld.himusic")
        .withAbilityName("com.domainname.hiworld.himusic.ServiceAbility")
        .build();
intent.setOperation(operation);
startAbility(intent);
```

启动远程设备 Service 的相关代码如下。

```
Intent intent = new Intent();
Operation operation = new Intent.OperationBuilder()
```

```
            .withDeviceId("deviceId")
            .withBundleName("com.domainname.hiworld.himusic")
            .withAbilityName("com.domainname.hiworld.himusic.ServiceAbility")
            .withFlags(Intent.FLAG_ABILITYSLICE_MULTI_DEVICE) //设置支持分布式调度系统多设备
                                                             //启动的标识
            .build();
    intent.setOperation(operation);
    startAbility(intent);
```

执行上述代码后,Ability 通过 startAbility()方法启动 Service。如果 Service 尚未运行,则系统会先调用 onStart()方法初始化 Service,再回调 Service 的 onCommand()方法启动 Service。如果 Service 正在运行,则系统会直接回调 Service 的 onCommand()方法启动 Service。

2) 停止 Service

Service 一旦创建就会一直保持在后台运行,除非必须回收内存资源,否则系统不会停止或销毁 Service。开发者可以在 Service 中通过 terminateAbility()停止本地 Service 或在其他 Ability 调用 stopAbility()停止 Service。

停止 Service 同样支持停止本地设备 Service 和停止远程设备 Service,使用方法与启动 Service 一样。一旦调用停止 Service 的方法,系统便会尽快销毁 Service。

4. 连接 Service

如果 Service 需要与 Page Ability 或其他应用的 Service Ability 进行交互,则需要创建用于连接的 Connection。Service 支持其他 Ability 通过 connectAbility()方法与其进行连接。

在使用 connectAbility()处理回调时,需要传入目标 Service 的 Intent 与 IAbilityConnection 的实例。IAbilityConnection 提供两种方法:onAbilityConnectDone()是处理连接 Service 成功的回调;onAbilityDisconnectDone()是处理 Service 异常死亡的回调,创建连接 Service 回调实例的相关代码如下。

```
private IAbilityConnection connection = new IAbilityConnection() {//连接到 Service 的回调
    @Override
    public void onAbilityConnectDone(ElementName elementName, IRemoteObject iRemoteObject,
int resultCode) {
        //Client 侧需要定义与 Service 侧相同的 IRemoteObject 实现类.开发者获取服务器端传输
        //的 IRemoteObject 对象,并从中解析出服务器端传输的信息
    }
    //Service 异常死亡的回调
    @Override
    public void onAbilityDisconnectDone(ElementName elementName, int resultCode) {
    }
};
```

连接 Service 相关代码如下。

```
//连接 Service
Intent intent = new Intent();
Operation operation = new Intent.OperationBuilder()
        .withDeviceId("deviceId")
        .withBundleName("com.domainname.hiworld.himusic")
        .withAbilityName("com.domainname.hiworld.himusic.ServiceAbility")
        .build();
intent.setOperation(operation);
connectAbility(intent, connection);
```

同时,Service 侧也需要在调用 onConnect()时返回 IRemoteObject,从而定义与 Service 进行通信的接口。onConnect()需要返回一个 IRemoteObject 对象,HarmonyOS 提供了 IRemoteObject 的默认实现,用户可以通过继承 LocalRemoteObject 创建自定义的实现类。Service 侧将自身的实例返回给调用侧的相关代码如下。

```
//创建自定义 IRemoteObject 实现类
private class MyRemoteObject extends LocalRemoteObject {
    MyRemoteObject(){
    }
}
//将 IRemoteObject 返回给客户端
@Override
protected IRemoteObject onConnect(Intent intent) {
    return new MyRemoteObject();
}
```

5. Service Ability 生命周期

与 Page 类似,Service 也拥有生命周期,如图 3-3 所示。根据调用方法不同,其生命周期有以下两种路径。

启动 Service,在其他 Ability 调用 startAbility()时创建,然后保持运行。其他 Ability 通过调用 stopAbility()停止 Service,Service 停止后,系统会将其销毁。

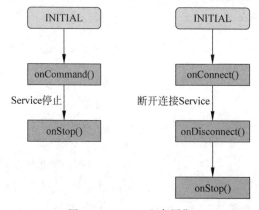

图 3-3　Service 生命周期

连接 Service，在其他 Ability 调用 connectAbility()时创建，客户端可通过调用 disconnectAbility()断开连接。多个客户端可以绑定相同的 Service，而且当所有绑定全部取消后，系统会销毁 Service。

connectAbility()也可以连接通过 startAbility()创建的 Service。

6. 前台 Service

一般情况下，Service 都是在后台运行的，后台 Service 的优先级比较低，当资源不足时，系统有可能回收正在运行的后台 Service。

在一些场景下(如播放音乐)，用户希望应用能够一直保持运行，此时就需要使用前台 Service，它始终保持正在运行的图标在系统状态栏显示。

使用前台 Service 并不复杂，只需在 Service 创建的方法中，调用 keepBackgroundRunning()将 Service 与通知绑定。调用 keepBackgroundRunning()方法前需要在配置文件中声明 ohos.permission.KEEP_BACKGROUND_RUNNING 权限，同时还需要在配置文件中添加对应的 backgroundModes 参数。在 onStop()方法中调用 cancelBackgroundRunning()方法可停止前台 Service，使用前台 Service 的 onStart()相关代码如下。

```
//创建通知，其中 1005 为 notificationId
NotificationRequest request = new NotificationRequest(1005);
NotificationRequest.NotificationNormalContent content = new NotificationRequest.NotificationNormalContent();
content.setTitle("title").setText("text");
NotificationRequest.NotificationContent notificationContent = new NotificationRequest.NotificationContent(content);
request.setContent(notificationContent);
//绑定通知,1005 为创建通知时传入的 notificationId
keepBackgroundRunning(1005, request);
```

在配置文件中，module > abilities 字段下对当前 Service 做如下配置。

```
{
    "name": ".ServiceAbility",
    "type": "service",
    "visible": true,
    "backgroundModes": ["dataTransfer", "location"]
}
```

针对 Service Ability 开发，启动、停止、连接、断开连接等支持对跨设备的 Service Ability 进行操作。示例工程参考地址为：https://gitee.com/openharmony/app_samples/tree/master/ability/ServiceAbility。

3.2.3　Data Ability

使用 Data 模板的 Ability(以下简称 Data)有助于应用管理其自身和其他应用存储数据的访问，并提供与其他应用共享数据的方法。Data 既可用于同设备不同应用的数据共享，也支持跨设备不同应用的数据共享。数据的存放形式多样，可以是数据库，也可以是磁盘上的文件。Data 对外提供对数据的增、删、改、查，以及打开文件等接口，这些接口的具体实现由开发者提供。

Data 的提供方和使用方都通过 URI(Uniform Resource Identifier)标识一个具体的数据，例如，数据库中的某个表或磁盘上的某个文件。HarmonyOS 的 URI 仍基于 URI 通用标准，格式如图 3-4 所示。

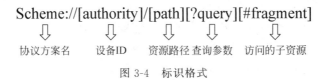

图 3-4　标识格式

Scheme：协议方案名，固定为 dataability，代表 Data Ability 所使用的协议类型。

authority：设备 ID，如果为跨设备场景，则为目标设备的 ID，如果为本地设备场景，则不需要填写。

path：资源的路径信息，代表特定资源的位置信息。

query：查询参数。

fragment：可以用于指示要访问的子资源。

URI 示例如下：

跨设备场景：dataability://device_id/com.domainname.dataability.persondata/person/10。

本地设备：dataability:///com.domainname.dataability.persondata/person/10。

本地设备的 device_id 字段为空，因此在 dataability：后面有三个"/"。

1. 创建 Data

使用 Data 模板的 Ability 形式仍然是 Ability，需要为应用添加一个或多个 Ability 的子类，提供程序与其他应用之间的接口。Data 为结构化数据和文件提供不同 API 接口供用户使用，因此需要确定好使用何种类型的数据。本部分主要讲述创建 Data 的基本步骤和需要使用的接口。

Data 提供方可以自定义数据的增、删、改、查，以及文件打开等功能，并对外提供这些接口。Data 支持以下两种数据形式：文件数据，例如文本、图片、音乐等，结构化数据，例如数据库等。

UserDataAbility 用于接收其他应用发送的请求，提供外部程序访问的入口，从而实现应用间的数据访问。实现 UserDataAbility，需要在 Project 窗口当前工程的主目录(entry→src→main→java→com.xxx.xxx)选择 File→New→Ability→Empty Data Ability，设置 Data Name 后完成 UserDataAbility 的创建。Data 提供了文件存储和数据库存储两组接口供用户使用。

1) 文件存储

开发者需要在 Data 中重写 FileDescriptor openFile(Uri uri, String mode)方法,操作如下:uri 为客户端传入的请求目标路径,mode 为开发者对文件的操作选项,可选方式包含 r(读),w(写),rw(读写)等。

通过 MessageParcel 静态方法 dupFileDescriptor()复制到操作文件流的文件描述符,并将其返回,供远端应用访问文件。根据传入的 uri 打开对应的文件代码如下。

```java
private static final HiLogLabel LABEL_LOG = new HiLogLabel(HiLog.LOG_APP, 0xD00201, "Data_Log");
@Override
public FileDescriptor openFile(Uri uri, String mode) throws FileNotFoundException {
    File file = new File(uri.getDecodedPathList().get(0)); //get(0)是获取 uri 中查询参数字段
    if (mode == null || !"rw".equals(mode)) {
        file.setReadOnly();
    }
    FileInputStream fileIs = new FileInputStream(file);
    FileDescriptor fd = null;
    try {
        fd = fileIs.getFD();
    } catch (IOException e) {
        HiLog.info(LABEL_LOG, "failed to getFD");
    }
    //绑定文件描述符
    return MessageParcel.dupFileDescriptor(fd);
}
```

2) 数据库存储

初始化数据库连接:系统会在应用启动时调用 onStart()方法创建 Data 实例。在此方法中,应该创建数据库连接,并获取连接对象,以便后续和数据库进行操作。为了避免影响应用启动速度,尽可能将非必要的耗时任务推迟到使用时执行,而不是在此方法中执行所有初始化,相关代码如下。

```java
private static final String DATABASE_NAME = "UserDataAbility.db";
private static final String DATABASE_NAME_ALIAS = "UserDataAbility";
private static final HiLogLabel LABEL_LOG = new HiLogLabel(HiLog.LOG_APP, 0xD00201, "Data_Log");
private OrmContext ormContext = null;
@Override
public void onStart(Intent intent) {
    super.onStart(intent);
    DatabaseHelper manager = new DatabaseHelper(this);
    ormContext = manager.getOrmContext(DATABASE_NAME_ALIAS, DATABASE_NAME, BookStore.class);
}
```

Ability 定义了 6 个方法供用户处理对数据库的增、删、改、查。这 6 个方法在 Ability 中已默认实现,开发者可按需重写,如表 3-1 所示。

表 3-1 编写数据库方法及功能描述

方 法	功 能 描 述
ResultSet query(Uri uri, String[] columns, DataAbilityPredicates predicates)	查询数据库
int insert(Uri uri, ValuesBucket value)	向数据库中插入单条数据
int batchInsert(Uri uri, ValuesBucket[] values)	向数据库中插入多条数据
int delete(Uri uri, DataAbilityPredicates predicates)	删除一条或多条数据
int update(Uri uri, ValuesBucket value, DataAbilityPredicates predicates)	更新数据库
DataAbilityResult[] executeBatch(ArrayList < DataAbilityOperation > operations)	批量操作数据库

在初始化数据库类 BookStore.class,并通过实体类 User.class 对该数据库的表 User 进行增、删、改、查操作,具体方法如下。

(1) query():接收三个参数,分别是查询的目标路径、列名及查询条件,查询条件由类 DataAbilityPredicates 构建。根据传入的列名和查询条件查询用户表,相关代码如下。

```
public ResultSet query(Uri uri, String[] columns, DataAbilityPredicates predicates) {
    if (ormContext == null) {
        HiLog.error(LABEL_LOG, "failed to query, ormContext is null");
        return null;
    }
    //查询数据库
    OrmPredicates ormPredicates = DataAbilityUtils.createOrmPredicates(predicates, User.class);
    ResultSet resultSet = ormContext.query(ormPredicates, columns);
    if (resultSet == null) {
        HiLog.info(LABEL_LOG, "resultSet is null");
    }
    //返回结果
    return resultSet;
}
```

(2) insert():接收两个参数,分别是插入的目标路径和数据值。其中,插入的数据由 ValuesBucket 封装,服务器端可以从该参数中解析出对应的属性,然后插入数据库中。此方法返回一个 int 类型的值用于标识结果。接收到传输过来的用户信息并保存到数据库中的相关代码如下。

```
public int insert(Uri uri, ValuesBucket value) {
    //参数校验
    if (ormContext == null) {
        HiLog.error(LABEL_LOG, "failed to insert, ormContext is null");
        return -1;
```

```java
        }
        //构造插入数据
        User user = new User();
        user.setUserId(value.getInteger("userId"));
        user.setFirstName(value.getString("firstName"));
        user.setLastName(value.getString("lastName"));
        user.setAge(value.getInteger("age"));
        user.setBalance(value.getDouble("balance"));
        //插入数据库
        boolean isSuccessful = ormContext.insert(user);
        if (!isSuccessful) {
            HiLog.error(LABEL_LOG, "failed to insert");
            return -1;
        }
        isSuccessful = ormContext.flush();
        if (!isSuccessful) {
            HiLog.error(LABEL_LOG, "failed to insert flush");
            return -1;
        }
        DataAbilityHelper.creator(this, uri).notifyChange(uri);
        int id = Math.toIntExact(user.getRowId());
        return id;
    }
```

(3) batchInsert()：为批量插入方法，接收一个 ValuesBucket 数组用于单次插入一组对象，它的作用是提高插入多条重复数据的效率。该方法系统已实现，可以直接调用。

(4) delete()：用来执行删除操作。删除条件由类 DataAbilityPredicates 构建，服务器端在接收到该参数之后可以从中解析出要删除的数据，然后到数据库中执行。根据传入的条件删除用户表数据的相关代码如下。

```java
    public int delete(Uri uri, DataAbilityPredicates predicates) {
        if (ormContext == null) {
            HiLog.error(LABEL_LOG, "failed to delete, ormContext is null");
            return -1;
        }
        OrmPredicates ormPredicates = DataAbilityUtils.createOrmPredicates(predicates, User.class);
        int value = ormContext.delete(ormPredicates);
        DataAbilityHelper.creator(this, uri).notifyChange(uri);
        return value;
    }
```

(5) update()：用来执行更新操作。用户可以在 ValuesBucket 参数中指定要更新的数据，在 DataAbilityPredicates 中构建更新的条件等，相关代码如下。

```java
    public int update(Uri uri, ValuesBucket value, DataAbilityPredicates predicates) {
```

```
    if (ormContext == null) {
        HiLog.error(LABEL_LOG, "failed to update, ormContext is null");
        return -1;
    }
    OrmPredicates ormPredicates = DataAbilityUtils.createOrmPredicates(predicates, User.class);
    int index = ormContext.update(ormPredicates, value);
    HiLog.info(LABEL_LOG, "UserDataAbility update value:" + index);
    DataAbilityHelper.creator(this, uri).notifyChange(uri);
    return index;
}
```

(6) executeBatch()：批量执行操作。DataAbilityOperation 中提供了设置操作类型、数据和操作条件的方法，用户可自行设置要执行的数据库操作。该方法系统已实现，可以直接调用。

2. 注册 UserDataAbility

与 Service 类似，在配置文件中必须注册 Data，该字段在创建 Data Ability 时会自动创建，name 与创建的 Data Ability 一致，需要关注以下属性。

Type：类型设置为 Data。URI：对外提供的访问路径，全局唯一。permissions：访问该 Data Ability 时需要申请的访问权限。如果是非系统权限，在配置文件中进行自定义，相关代码如下。

```
{
    "name": ".UserDataAbility",
    "type": "data",
    "visible": true,
    "uri": "dataability://com.example.myapplication5.DataAbilityTest",
    "permissions": [
        "com.example.myapplication5.DataAbility.DATA"
    ]
}
```

3. 访问 Data

通过 DataAbilityHelper 访问当前应用或其他应用提供的共享数据。DataAbilityHelper 作为客户端，与提供方的 Data 进行通信。Data 接收到请求后，执行相应的处理，并返回结果。DataAbilityHelper 提供了一系列与 Data Ability 对应的方法，使用步骤如下。

如果待访问的 Data 声明了访问需要权限，则访问此 Data 需要在配置文件中进行声明。

```
"reqPermissions": [
    {
        "name": "com.example.myapplication5.DataAbility.DATA"
    },
    //访问文件需要添加访问存储读写权限
```

```
    {
        "name": "ohos.permission.READ_USER_STORAGE"
    },
    {
        "name": "ohos.permission.WRITE_USER_STORAGE"
    }
]
```

创建 DataAbilityHelper：DataAbilityHelper 为开发者提供 creator（）方法创建 DataAbilityHelper 实例。此方法为静态方法，有多个重载。最常见的方法是通过传入一个 context 对象创建 DataAbilityHelper 对象，获取 Helper 对象示例如下。

```
DataAbilityHelper helper = DataAbilityHelper.creator(this)
```

访问 Data Ability：DataAbilityHelper 为开发者提供一系列的接口，访问不同类型的数据(文件、数据库等)。

访问文件：DataAbilityHelper 为开发者提供 FileDescriptor openFile(Uri uri, String mode)方法操作文件。此方法需要传入两个参数，其中 URI 确定目标资源路径，Mode 指定打开文件的方式，可选方式包含 r(读)、w(写)、rw(读写)、wt(覆盖写)、wa(追加写)、rwt(覆盖写且可读)。

此方法返回一个目标文件的 FD(文件描述符)，把文件描述符封装成流，就可以对文件流进行自定义处理，访问文件示例如下。

```
FileDescriptor fd = helper.openFile(uri, "r");   //读取文件描述符
FileInputStream fis = new FileInputStream(fd); //使用文件描述符封装成的文件流,进行文件操作
```

访问数据库：DataAbilityHelper 提供增、删、改、查及批量处理等方法操作数据库。

针对 Data Ability 开发，示例工程参考地址为：https://gitee.com/openharmony/app_samples/tree/master/ability/DataAbility。

3.2.4 Intent

Intent 是对象之间传递信息的载体。例如，当一个 Ability 需要启动另一个 Ability 时，或者一个 AbilitySlice 需要导航到另一个 AbilitySlice 时，可以通过 Intent 指定启动的目标同时携带相关数据。Intent 的构成元素包括 Operation 与 Parameters，如表 3-2 所示。

Intent 用于发起请求时，根据指定元素的不同，操作分为两种类型：如果同时指定了 BundleName 与 AbilityName，则根据 Ability 的全称(如 com.demoapp.FooAbility)直接启动应用；如果未同时指定 BundleName 和 AbilityName，则根据 Operation 中的其他属性启动应用。Intent 设置属性时，必须先使用 Operation 进行设置。如果需要新增或修改属性，必须在设置 Operation 后再执行操作。

表 3-2　Operation 与 Parameters 描述

属性	子属性	功能描述
Operation	Action	表示动作，通常使用系统预置 Action，应用也可以自定义 Action，例如 IntentConstants.ACTION_HOME 表示返回桌面动作
	Entity	表示类别，通常使用系统预置 Entity，应用也可以自定义 Entity，例如 Intent.ENTITY_HOME 表示在桌面显示图标
	URI	表示 URI 描述。如果在 Intent 中指定了 URI，则 Intent 将匹配指定的 URI 信息，包括 scheme、schemeSpecificPart、authority 和 path 信息
	Flags	表示处理 Intent 的方式，例如 Intent.FLAG_ABILITY_CONTINUATION 标记在本地的一个 Ability 是否可以迁移到远端设备继续运行
	BundleName	表示包描述。如果在 Intent 中同时指定 BundleName 和 AbilityName，则 Intent 可以直接匹配到指定的 Ability
	AbilityName	表示待启动的 Ability 名称。如果在 Intent 中同时指定 BundleName 和 AbilityName，则 Intent 可以直接匹配到指定的 Ability
	DeviceId	表示运行指定 Ability 的设备 ID
Parameters	—	Parameters 是一种支持自定义的数据结构，开发者可以通过 Parameters 传递某些请求所需的额外信息

1. 根据 Ability 的全称启动应用

通过构造包含 BundleName 与 AbilityName 的 Operation 对象，可以启动一个 Ability，并导航到该 Ability，相关代码如下。

```
Intent intent = new Intent();
//通过 Intent 中的 OperationBuilder 类构造 operation 对象，指定设备标识(空串表示当前设备)、
//应用包名、Ability 名称
Operation operation = new Intent.OperationBuilder()
        .withDeviceId("")
        .withBundleName("com.demoapp")
        .withAbilityName("com.demoapp.FooAbility")
        .build();
//将 operation 设置到 Intent 中
intent.setOperation(operation);
startAbility(intent);
```

Intent 作为处理请求的对象，会在相应的回调方法中接收请求方传递的 Intent 对象。以导航到另一个 Ability 为例，导航的目标 Ability 可以在 onStart()回调的参数中获得 Intent 对象。

2. 根据 Operation 的其他属性启动应用

有些场景下需要使用其他应用提供的某种能力，例如通过浏览器打开一个链接，而不关

心用户最终选择哪个浏览器应用,则可以通过 Operation 的其他属性(除 BundleName 与 AbilityName 之外的属性)描述需要的能力。如果设备上存在多个应用提供同种能力,系统则弹出候选列表,由用户选择使用哪个应用处理请求。以下示例展示使用 Intent 跨 Ability 查询天气信息。

1)请求方

在 Ability 中构造 Intent 及包含 Action 的 Operation 对象,并调用 startAbilityForResult() 方法发起请求,然后重写 onAbilityResult() 回调方法,对请求结果进行处理,相关代码如下。

```java
private void queryWeather() {
    Intent intent = new Intent();
    Operation operation = new Intent.OperationBuilder()
            .withAction(Intent.ACTION_QUERY_WEATHER)
            .build();
    intent.setOperation(operation);
    startAbilityForResult(intent, REQ_CODE_QUERY_WEATHER);
}
@Override
protected void onAbilityResult(int requestCode, int resultCode, Intent resultData) {
    switch (requestCode) {
        case REQ_CODE_QUERY_WEATHER:
            //处理结果
            ...
            return;
        default:
            ...
    }
}
```

2)处理方

作为处理请求对象,需要在配置文件中声明对外提供的能力,以便系统可以找到并进行处理,相关代码如下。

```
{
    "module": {
        ...
        "abilities": [
            {
                ...
                "skills":[
                    {
                        "actions":[
                            "ability.intent.QUERY_WEATHER"
```

```
                    ]
                }
            ]
            ...
            }
        ]
        ...
    }
    ...
}
```

在 Ability 中配置路由以便支持以 action 导航到对应的 AbilitySlice,相关代码如下。

```
@Override
protected void onStart(Intent intent) {
    ...
    addActionRoute(Intent.ACTION_QUERY_WEATHER, DemoSlice.class.getName());
    ...
}
```

在 Ability 中处理请求,并调用 setResult()方法暂存返回结果,相关代码如下。

```
@Override
protected void onActive() {
    ...
    Intent resultIntent = new Intent();
    setResult(0, resultIntent);    //0 为当前 Ability 销毁后返回的 resultCode
    ...
}
```

针对 Intent 开发,演示如何根据 Ability 全称启动应用和根据 Operation 的其他属性启动示例工程,参考地址为: https://gitee.com/openharmony/app_samples/tree/master/ability/Intent。

3.2.5 Ability 示例

在 HarmonyOS 中,Ability 是应用所具备能力的抽象,一个应用可以具备多种能力(包含多个 Ability),HarmonyOS 支持应用以 Ability 为单位进行部署。其中,Page Ability 提供与用户交互的能力,一个 Page 可以由一个或多个 AbilitySlice 构成,AbilitySlice 指应用的单个页面及其控制逻辑的总和。示例工程参考地址为:https://developer.huawei.com/consumer/cn/codelabsPortal/carddetails/Ability_Intent。

在一个 Ability 中,可以使用 present()/presentForResult() 方法,从一个 AbilitySlice 导航到一个新的 AbilitySlice,即 Page 内不同 AbilitySlice 的跳转。同时,也可以在 AbilitySlice 中使用 startAbility()/startAbilityForResult() 方法启动一个新的 Ability。

在本示例中,尝试构建 2 个 Ability、3 个 AbilitySlice 完成两种类型的跳转。在 MainAbilitySlice 页面,写两个简单的按钮,其中一个实现 Page 内跳转,如图 3-5 所示,另一个实现 Page 间跳转,如图 3-6 所示。在 NewAbilitySlice 和 SecondAbilitySlice 页面,通过一个按钮让它回到第一个页面,单击第一个按钮跳转到第二个 AbilitySlice,再跳回;单击第二个按钮跳转到第二个 Ability,再跳回。

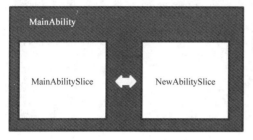

图 3-5　Page 内跳转

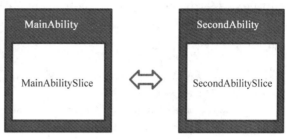

图 3-6　Page 间跳转

如图 3-7 所示,构造 Operation 对象,通过 startAbility()/startAbilityForResult()实现不同 Page 间导航。

图 3-7　Page 间导航

1. 新建项目

打开 DevEco Studio 开发环境,以使用 Java 语言开发、设备类型为 Phone 的 Application 为例,模板选择 Empty Feature Ability(Java)。工程创建完成后,使用 Phone 模拟器运行

工程。

2. 编写页面布局

在 Java UI 框架中，系统提供两种编写布局的方式：UI 布局和在代码中创建布局。以 XML 方式为例，在 XML 中声明 UI 布局和在代码中创建布局步骤如下。

在创建项目后，系统自动创建 MainAbility 和 MainAbilitySlice。在 Project 窗口单击 entry→src→main→resources→base→layout，打开 ability_main.xml 文件。

编写一个文本和两个按钮，使用 DependentLayout 布局，通过 Text 和 Button 组件实现，ability_main.xml 的相关代码如下。

```xml
<?xml version = "1.0" encoding = "utf-8"?>
<DirectionalLayout
    xmlns:ohos = "http://schemas.huawei.com/res/ohos"
    ohos:height = "match_parent"
    ohos:width = "match_parent"
    ohos:alignment = "center"
    ohos:orientation = "vertical">
    <Text
        ohos:id = "$+id:main_text"
        ohos:text = "嗨,我是 MainAbilitySlice,看标题就知道我在哪个 Page 啦"
        ohos:width = "320fp"
        ohos:height = "match_content"
        ohos:text_alignment = "center"
        ohos:multiple_lines = "true"
        ohos:text_size = "20fp"
        ohos:above = "$+id:enter_second"
        ohos:bottom_margin = "50vp"/>
    <Button
        ohos:id = "$+id:enter_newAbilitySlice"
        ohos:width = "260vp"
        ohos:height = "45vp"
        ohos:text = "导航到 NewAbilitySlice"
        ohos:multiple_lines = "true"
        ohos:text_size = "20fp"
        ohos:bottom_margin = "20vp"
        ohos:center_in_parent = "true"
        ohos:background_element = "$graphic:background_ability_main"/>
    <Button
        ohos:id = "$+id:enter_second"
        ohos:width = "260vp"
        ohos:height = "45vp"
        ohos:text = "导航到 SecondAbility"
        ohos:multiple_lines = "true"
        ohos:text_size = "20fp"
        ohos:bottom_margin = "20vp"
        ohos:center_in_parent = "true"
```

```xml
        ohos:background_element = " $graphic:background_ability_main"/>
</DirectionalLayout >
```

代码中的按钮使用 background_element＝" $graphic：background_ability_main " ，即 entry→src→main→resources→base→graphic 中的 background_ability_main. xml 文件，完成上述步骤后，实现第一个页面布局，相关代码如下。

```xml
< shape xmlns:ohos = "http://schemas.huawei.com/res/ohos"
        ohos:shape = "rectangle">
    < corners
            ohos:radius = "10"/>
    < solid
            ohos:color = "#ffc0c0c0"/>
    < stroke
            ohos:color = "#ff00ff7f"
            ohos:width = "0.4vp"/>
</shape >
```

创建第二个 Page，在 Project 窗口中，打开 entry→src→main→java→包名，右击 slice 文件夹，选择 New→Ability→Empty Page Ability(Java)，命名为 SecondAbility，按下 Enter 键，系统会创建 SecondAbility. java 和 slice/SecondAbilitySlice. java。

首先，在 Project 窗口中，单击 entry→src→main→resources→base→layout，打开 ability_second. xml 文件，增加信息显示区域和跳转按钮，完成第二个页面布局，相关代码如下。

```xml
<?xml version = "1.0" encoding = "utf-8"?>
< DirectionalLayout
    xmlns:ohos = "http://schemas.huawei.com/res/ohos"
    ohos:height = "match_parent"
    ohos:width = "match_parent"
    ohos:alignment = "center"
    ohos:orientation = "vertical">
    < Text
        ohos:id = " $ + id:second_text"
        ohos:width = "320fp"
        ohos:height = "match_content"
        ohos:multiple_lines = "true"
        ohos:text = "嗨, 我是 SecondAbilitySlice"
        ohos:text_alignment = "center"
        ohos:text_size = "20fp"
        ohos:above = " $ + id:second_back_first"
        ohos:bottom_margin = "50vp"/>
    < Button
        ohos:id = " $ + id:second_back_first"
        ohos:width = "260vp"
        ohos:height = "45vp"
```

```xml
            ohos:text = "回到 MainAbility"
            ohos:multiple_lines = "true"
            ohos:text_size = "20fp"
            ohos:bottom_margin = "20vp"
            ohos:center_in_parent = "true"
            ohos:background_element = " $graphic:background_ability_main"/>
</DirectionalLayout>
```

其次,在 entry→src→main→java→包名→slice 文件夹上右击新建一个 NewAbilitySlice. java 文件,并在 entry→src→main→resources→base→layout 中新建 ability_main_new. xml 文件,用于实现 Page 内跳转调用。在 ability_main_new. xml 文件中,同样增加一个信息显示区域和跳转按钮,相关代码如下。

```xml
<?xml version = "1.0" encoding = "utf - 8"?>
<DirectionalLayout
    xmlns:ohos = "http://schemas.huawei.com/res/ohos"
    ohos:height = "match_parent"
    ohos:width = "match_parent"
    ohos:alignment = "center"
    ohos:orientation = "vertical">
    <Text
        ohos:id = " $ + id:new_text"
        ohos:width = "320fp"
        ohos:height = "match_content"
        ohos:multiple_lines = "true"
        ohos:text = "我是 NewAbilitySlice.麻烦看看标题,请告诉我现在在哪个 Page 呢?"
        ohos:text_alignment = "center"
        ohos:text_size = "20fp"
        ohos:above = " $ + id:second_back_first"
        ohos:bottom_margin = "50vp"/>
    <Button
        ohos:id = " $ + id:new_to_main"
        ohos:width = "260vp"
        ohos:height = "45vp"
        ohos:text = "回到 MainAbilitySlice"
        ohos:multiple_lines = "true"
        ohos:text_size = "20fp"
        ohos:bottom_margin = "20vp"
        ohos:center_in_parent = "true"
        ohos:background_element = " $graphic:background_ability_main"/>
</DirectionalLayout>
```

最后,对 Ability 页面的标题栏进行修改,便于分辨页面。在 Project 窗口中,单击 entry→src→main,打开 config.json 文件,修改 abilities 中两个 Label 字段分别为 Page MainAbility 和 Page SecondAbility。默认情况下,Label 字段生成为引用模式不可直接修改,按住 Ctrl 键单击该字段,跳转到被引用的 string.json 文件,修改对应字段即可,相关代码如下。

```
{
  "string": [
    {
      "name": "entry_MainAbility",
      "value": "Page MainAbility"
    },
    {
      "name": "entry_SecondAbility",
      "value": "Page SecondAbility"
    },
    ...
  ]
}
```

3. 编写 AbilitySlice 间的导航

打开 entry→src→main→java→包名→slice 中的 MainAbilitySlice.java 文件,添加两个按钮的响应逻辑,实现单击按钮跳转到下一页。

其中,Page 内 AbilitySlice 间的导航选择使用 presentForResult()实现,可以获得从导航目标 AbilitySlice 返回时的结果,相关代码如下。

```java
private Text backValueText;
@Override
public void onStart(Intent intent) {
    super.onStart(intent);
    super.setUIContent(ResourceTable.Layout_ability_main);
    //开始进入 NewAbilitySlice
    Component enterNewAbilitySliceButton = findComponentById(ResourceTable.Id_enter_newAbilitySlice);
    enterNewAbilitySliceButton.setClickedListener(listener -> presentForResult(new NewAbilitySlice(), new Intent(), 0));
    //开始进入 SecondAbility page
    Component enterSecondAbilityButton = findComponentById(ResourceTable.Id_enter_second);
    enterSecondAbilityButton.setClickedListener(component -> startEnterSecondAbility());
    backValueText = (Text) findComponentById(ResourceTable.Id_main_text);
}
```

打开同文件夹下新建的 NewAbilitySlice.Java 文件,通过 setResult()完成返回结果的实现,相关代码如下。

```java
package com.huawei.abilityintent.slice;
import com.huawei.abilityintent.ResourceTable;
import ohos.aafwk.ability.AbilitySlice;
import ohos.aafwk.content.Intent;
import ohos.agp.components.Component;
public class NewAbilitySlice extends AbilitySlice {
```

```java
    @Override
    public void onStart(Intent intent) {
        super.onStart(intent);
        super.setUIContent(ResourceTable.Layout_ability_main_new);
        Component newToMainButton = findComponentById(ResourceTable.Id_new_to_main);
        newToMainButton.setClickedListener(component -> terminate());
    }
    @Override
    public void onActive() {
        super.onActive();
        Intent intent = new Intent();
        intent.setParam("key", "我从NewAbilitySlice跳回来咯");
        setResult(intent);
    }
    @Override
    public void onForeground(Intent intent) {
        super.onForeground(intent);
    }
}
```

继续在 MainAbilitySlice.java 中补充如下代码，通过 onResult 完成返回结果的接收，并显示在页面的 Text 控件中，相关代码如下。

```java
/* presentForResult()结果 */
@Override
protected void onResult(int requestCode, Intent resultIntent) {
    if (requestCode != 0 || resultIntent == null) {
        return;
    }
    String result = resultIntent.getStringParam("key");
    backValueText.setText(result);
}
```

4. 编写 Page 之间的导航

在 entry→src→main→java→包名的 MainAbility.java 文件内，可以看到在 onStart() 中已经通过路由导航到 MainAbilitySlice 页面。

```java
public class MainAbility extends Ability {
    @Override
    public void onStart(Intent intent) {
        super.onStart(intent);
        super.setMainRoute(MainAbilitySlice.class.getName());
    }
}
```

同理，SecondAbility.java 也已经导航到 SecondAbilitySlice.java 页面，接下来完成两

个 Page Ability 之间的导航。

打开 entry→src→main→java→包名→slice 中的 MainAbilitySlice.java 文件,进一步完成 Page 间导航按钮的响应逻辑,实现对应的方法 startEnterSecondAbility()。

Page 间的导航可以使用 startAbility()或 startAbilityForResult()方法,获得返回结果的回调为 onAbilityResult()。为了获取返回值,这里使用 startAbilityForResult()方法跳转到 SecondAbility,相关代码如下。

```java
/*显式启动*/
private void startEnterSecondAbility() {
    Intent intent = new Intent();
    Operation operation = new Intent.OperationBuilder().withDeviceId("")
            .withBundleName(getBundleName())
            .withAbilityName("com.huawei.abilityintent.SecondAbility")
            .build();
    intent.setOperation(operation);
    intent.setParam("key", "我从MainAbility进到了SecondAbility");
    startAbilityForResult(intent, 1);
}
```

在同目录下的 SecondAbilitySlice.java 文件中,对从 MainAbility 中获取到的信息通过 Text 控件进行展示,并且通过按钮单击事件完成 terminate()操作,相关代码如下。

```java
@Override
public void onStart(Intent intent) {
    super.onStart(intent);
    super.setUIContent(ResourceTable.Layout_ability_second);
    Component secondBackFirstButton = findComponentById(ResourceTable.Id_second_back_first);
    secondBackFirstButton.setClickedListener(component -> terminate());
    Text showParametersText = (Text) findComponentById(ResourceTable.Id_second_text);
    showParametersText.setText(intent.getStringParam("key"));
}
```

同时,在 SecondAbility 页面中设置返回 MainAbility 需要的数据。在 SecondAbility.java 中,调用 setResult()设置返回结果,相关代码如下。

```java
@Override
protected void onActive() {
    super.onActive();
    Intent intent = new Intent();
    intent.setParam("key", "我从SecondAbility跳回来啦");
    setResult(0, intent);
}
```

如果是在 AbilitySlice 中,也可以通过如下方式完成 setResult()设置返回结果,即 getAbility().setResult(0,intent)。

在 MainAbilitySlice.java 中增加如下代码,完成返回结果的回调函数 onAbilityResult(),将返回结果显示在页面上。

```
@Override
protected void onAbilityResult(int requestCode, int resultCode, Intent resultData) {
    if (resultCode != 0 || resultData == null) {
        return;
    }
    String result = resultData.getStringParam("key");
    backValueText.setText(result);
}
```

总之,Page 内可以使用 present()或 presentForResult()实现导航,使用 presentForResult()时可以获得从导航目标 AbilitySlice 返回时的返回结果。返回结果由导航目标 AbilitySlice 在其生命周期内通过 setResult()进行设置。当用户从导航目标 AbilitySlice 返回时,系统将回调 onResult()接收和处理返回结果,需要重写该方法。

Page 间的导航可以使用 startAbility() 或 startAbilityForResult() 方法,使用 startAbilityForResult()方法时可以获得导航目标 Ability 返回时的返回结果。在导航目标 Ability 中调用 setResult()设置返回结果,获得返回结果的回调为 onAbilityResult()。

另外一个实用示例是跨设备迁移支持将 Page 在同一用户的不同设备间迁移,以便支持用户无缝切换的需求。通过两个页面、两台设备实现在不同设备上的迁移。

3.3 公共事件与通知开发

HarmonyOS 通过 CES(Common Event Service,公共事件服务)为应用程序提供订阅、发布、退订公共事件的能力,通过 ANS(Advanced Notification Service,通知增强服务)系统服务为应用程序提供发布通知的能力。

公共事件可分为系统公共事件和自定义公共事件。系统公共事件:将收集到的事件信息,根据系统策略发送给订阅该事件的用户程序。包括如下内容:终端设备用户可感知的亮灭屏事件,以及系统关键服务发布的系统事件(如 USB 插拔、网络连接、系统升级)等。自定义公共事件:应用自定义一些公共事件处理业务逻辑。

通知提供应用的即时/通信消息,用户可以直接删除或单击通知触发进一步的操作。IntentAgent 封装了一个指定行为的 Intent,可以通过 IntentAgent 启动 Ability 和发布公共事件。应用如果需要接收公共事件,需要订阅相应的事件。

目前,公共事件与通知开发具有一定的约束与限制,具体内容如下。

公共事件的约束与限制:目前公共事件仅支持动态订阅,不支持多用户,部分系统事件需要具有指定的权限。ThreadMode 表示线程模型,目前仅支持 HANDLER 模式,即在当前 UI 线程上执行回调函数。DeviceId 用来指定订阅本地公共事件还是远端公共事件。DeviceId 为 Null、空字符串或本地设备 DeviceId 时,表示订阅本地公共事件,否则表示订阅远端公共事件。

通知的约束与限制：通知目前支持 6 种样式：普通文本、长文本、图片、社交、多行文本和媒体样式。创建通知时必须包含一种样式。注意：通知支持快捷回复。

IntentAgent 的限制：使用 IntentAgent 启动 Ability 时，Intent 必须指定 Ability 的包名和类名。

3.3.1 公共事件开发

每个应用都可以订阅自己感兴趣的公共事件，订阅成功且公共事件发布后，系统会将其发送给应用。这些公共事件可能来自系统、其他应用和应用自身。HarmonyOS 提供了一套完整的 API，支持用户订阅、发布和接收公共事件。发布公共事件需要借助 CommonEventData 对象，接收公共事件需要继承 CommonEventSubscriber 类并实现 onReceiveEvent 回调函数。

1. 公共事件开发接口关系

公共事件相关基础类包含 CommonEventData、CommonEventPublishInfo、CommonEventSubscribeInfo、CommonEventSubscriber 和 CommonEventManager，公共事件基础类关系如图 3-8 所示。

图 3-8　公共事件基础类关系

1) CommonEventData

CommonEventData 封装公共事件相关信息。用于在发布、分发和接收时处理数据。在构造 CommonEventData 对象时，相关参数需要注意以下事项：Code 为有序公共事件的结果码，Data 为有序公共事件的结果数据，仅用于有序公共事件场景。Intent 不允许为空，否则发布公共事件失败。主要接口及功能描述如表 3-3 所示。

表 3-3　CommonEventData 的主要接口及功能描述

主要接口	功能描述
CommonEventData()	创建公共事件数据
CommonEventData(Intent intent)	创建公共事件数据指定 Intent
CommonEventData(Intent intent, int code, String data)	创建公共事件数据，指定 Intent、Code 和 Data

续表

主要接口	功能描述
getIntent()	获取公共事件 Intent
setCode(int code)	设置有序公共事件的结果码
getCode()	获取有序公共事件的结果码
setData(String data)	设置有序公共事件的详细结果数据
getData()	获取有序公共事件的详细结果数据

2) CommonEventPublishInfo

CommonEventPublishInfo 封装公共事件发布相关属性、限制等信息，包括公共事件类型(有序或黏性)、接收者权限等。

有序公共事件：主要场景是多个订阅者有依赖关系或者对处理顺序有要求，例如高优先级订阅者可修改公共事件内容或处理结果，包括终止公共事件处理；或者低优先级订阅者依赖高优先级的处理结果等。订阅者可以通过 CommonEventSubscribeInfo.setPriority()方法指定优先级，默认为 0，优先级范围为[-1000,1000]，值越大优先级越高。

黏性公共事件：指公共事件的订阅动作在公共事件发布之后进行，订阅者也能收到公共事件类型。主要场景是由公共事件服务记录某些系统状态，例如蓝牙、WLAN、充电等事件和状态。不使用黏性公共事件机制时，应用可以通过直接访问系统服务获取该状态；在状态变化时，系统服务、硬件需要提供类似 observer 等方式通知应用。发布黏性公共事件可以通过 setSticky()方法设置，发布黏性公共事件需要申请如下权限。在 config.json 文件中的 reqPermissions 实现权限申请，字段说明如下。

```
"reqPermissions": [
  {
    "name": "ohos.permission.COMMONEVENT_STICKY",
    "reason": "Obtain the required permission",
    "usedScene": {
      "ability": [
        ".MainAbility"
      ],
      "when": "inuse"
    }
  },
  {
    ...
  }
]
```

CommonEventPublishInfo 主要接口及功能描述如表 3-4 所示。

表 3-4 CommonEventPublishInfo 主要接口及功能描述

主要接口	功能描述
CommonEventPublishInfo()	创建公共事件信息
CommonEventPublishInfo(CommonEventPublishInfo publishInfo)	复制公共事件信息
setSticky(boolean sticky)	设置公共事件的黏性属性
setOrdered(boolean ordered)	设置公共事件的有序属性
setSubscriberPermissions(String[] subscriberPermissions)	设置公共事件订阅者的权限,多参数仅第一个生效

3) CommonEventSubscribeInfo

CommonEventSubscribeInfo 封装公共事件订阅相关信息,例如优先级、线程模式、事件范围等。

线程模式(ThreadMode):设置订阅者的回调方法执行的线程模式。ThreadMode 有 HANDLER、POST、ASYNC 和 BACKGROUND 4 种模式,目前只支持 HANDLER 模式。

HANDLER:在 Ability 的主线程执行。

POST:在事件分发线程执行。

ASYNC:在一个新创建的异步线程执行。

BACKGROUND:在后台线程执行。

主要接口及功能描述如表 3-5 所示。

表 3-5 CommonEventSubscribeInfo 的主要接口及功能描述

主要接口	功能描述
CommonEventSubscribeInfo(MatchingSkills matchingSkills)	创建公共事件订阅器,指定 matchingSkills
CommonEventSubscribeInfo(CommonEventSubscribeInfo)	复制公共事件订阅器对象
setPriority(int priority)	设置优先级,用于有序公共事件
setThreadMode(ThreadMode threadMode)	指定订阅者回调函数运行在哪个线程上
setPermission(String permission)	设置发布者必须具备的权限
setDeviceId(String deviceId)	指定订阅哪台设备的公共事件

4) CommonEventSubscriber

CommonEventSubscriber.AsyncCommonEventResult 类处理有序公共事件异步执行。目前只能通过调用 CommonEventManager 的 subscribeCommonEvent() 进行订阅。主要接口及功能描述如表 3-6 所示。

表 3-6 CommonEventSubscriber 的主要接口及功能描述

主要接口	功能描述
CommonEventSubscriber(CommonEventSubscribeInfo subscribeInfo)	构造公共事件订阅者实例
onReceiveEvent(CommonEventData data)	由开发者实现,在接收到公共事件时被调用

续表

主要接口	功能描述
AsyncCommonEventResult goAsyncCommonEvent()	设置有序公共事件异步执行
setCodeAndData(int code,String data)	设置有序公共事件异步结果
setData(String data)	设置有序公共事件异步结果数据
setCode(int code)	设置有序公共事件异步结果码
getData()	获取有序公共事件异步结果数据
getCode()	获取有序公共事件异步结果码
abortCommonEvent()	取消当前公共事件,仅对有序公共事件有效,取消后,公共事件不再向下一个订阅者传递
getAbortCommonEvent()	获取当前有序公共事件是否取消的状态
clearAbortCommonEvent()	清除当前有序公共事件 Abort 状态
isOrderedCommonEvent()	查询当前公共事件是否为有序公共事件
isStickyCommonEvent()	查询当前公共事件是否为黏性公共事件

5) CommonEventManager

CommonEventManager 是为应用提供订阅、退订和发布公共事件的静态接口类,方法及功能描述如表 3-7 所示。

表 3-7　CommonEventManager 的方法及功能描述

方法	功能描述
publishCommonEvent(CommonEventData eventData)	发布公共事件
publishCommonEvent(CommonEventData event, CommonEventPublishInfo publishInfo)	发布公共事件,指定发布信息
publishCommonEvent(CommonEventData event, CommonEventPublishInfo publishInfo,CommonEventSubscriber resultSubscriber)	发布有序公共事件,指定发布信息和最后一个接收者
subscribeCommonEvent(CommonEventSubscriber subscriber)	订阅公共事件
unsubscribeCommonEvent(CommonEventSubscriber subscriber)	退订公共事件

2. 发布公共事件

开发者可以发布 4 种公共事件:无序公共事件、携带权限公共事件、有序公共事件和黏性公共事件。

1) 发布无序公共事件

构造 CommonEventData 对象,设置 Intent,通过构造 operation 对象把需要发布的公共事件信息传入 Intent 对象。

调用 CommonEventManager.publishCommonEvent(CommonEventData) 接口发布公共事件,相关代码如下。

```
try {
    Intent intent = new Intent();
    Operation operation = new Intent.OperationBuilder()
```

```java
            .withAction("com.my.test")
            .build();
    intent.setOperation(operation);
    CommonEventData eventData = new CommonEventData(intent);
    CommonEventManager.publishCommonEvent(eventData);
    HiLog.info(LABEL_LOG, "Publish succeeded");
} catch (RemoteException e) {
    HiLog.error(LABEL_LOG, "Exception occurred during publishCommonEvent invocation.");
}
```

2）发布携带权限公共事件

构造 CommonEventPublishInfo 对象，设置订阅者的权限。订阅者在 config.json 中申请所需的权限，发布带权限公共事件相关代码如下。

```java
Intent intent = new Intent();
Operation operation = new Intent.OperationBuilder()
        .withAction("com.my.test")
        .build();
intent.setOperation(operation);
CommonEventData eventData = new CommonEventData(intent);
CommonEventPublishInfo publishInfo = new CommonEventPublishInfo();
String[] permissions = {"com.example.MyApplication.permission"};
publishInfo.setSubscriberPermissions(permissions); //设置权限
try {
    CommonEventManager.publishCommonEvent(eventData, publishInfo);
    HiLog.info(LABEL_LOG, "Publish succeeded");
} catch (RemoteException e) {
    HiLog.error(LABEL_LOG, "Exception occurred during publishCommonEvent invocation.");
}
```

3）发布有序公共事件

构造 CommonEventPublishInfo 对象，通过 setOrdered(true) 指定公共事件属性为有序公共事件，也可以指定一个最后的公共事件接收者，相关代码如下。

```java
CommonEventSubscriber resultSubscriber = new MyCommonEventSubscriber();
CommonEventPublishInfo publishInfo = new CommonEventPublishInfo();
publishInfo.setOrdered(true); //设置属性为有序公共事件
try {
    CommonEventManager.publishCommonEvent(eventData, publishInfo, resultSubscriber);
                        //指定 resultSubscriber 为有序公共事件最后一个接收者
} catch (RemoteException e) {
    HiLog.error(LABEL_LOG, "Exception occurred during publishCommonEvent invocation.");
}
```

4）发布黏性公共事件

构造 CommonEventPublishInfo 对象，通过 setSticky(true) 指定公共事件属性为黏性

公共事件。发布者在 config.json 中申请发布黏性公共事件所需的权限,相关代码如下。

```
CommonEventPublishInfo publishInfo = new CommonEventPublishInfo();
publishInfo.setSticky(true); //设置属性为黏性公共事件
try {
    CommonEventManager.publishCommonEvent(eventData, publishInfo);
} catch (RemoteException e) {
    HiLog.error(LABEL, "Exception occurred during publishCommonEvent invocation.");
}
```

3. 订阅公共事件

订阅公共事件步骤如下。

(1) 创建 CommonEventSubscriber 派生类,在 onReceiveEvent() 回调函数中处理公共事件。此处不能执行耗时操作,否则会阻塞 UI 线程,产生用户单击没有反应等异常。

```
class MyCommonEventSubscriber extends CommonEventSubscriber {
    MyCommonEventSubscriber(CommonEventSubscribeInfo info) {
        super(info);
    }
    @Override
    public void onReceiveEvent(CommonEventData commonEventData) {
    }
}
```

(2) 构造 MyCommonEventSubscriber 对象,调用 CommonEventManager.subscribeCommonEvent() 接口进行订阅。

```
String event = "com.my.test";
MatchingSkills matchingSkills = new MatchingSkills();
matchingSkills.addEvent(event); //自定义事件
matchingSkills.addEvent(CommonEventSupport.COMMON_EVENT_SCREEN_ON);
//亮屏事件
CommonEventSubscribeInfo subscribeInfo = new CommonEventSubscribeInfo(matchingSkills);
MyCommonEventSubscriber subscriber = new MyCommonEventSubscriber(subscribeInfo);
try {
    CommonEventManager.subscribeCommonEvent(subscriber);
} catch (RemoteException e) {
    HiLog.error(LABEL, "Exception occurred during subscribeCommonEvent invocation.");
}
```

如果订阅拥有指定权限应用发布的公共事件,发布者需要在 config.json 中申请权限。如果订阅的公共事件是有序的,可以调用 setPriority() 指定优先级,相关代码如下。

```
String event = "com.my.test";
MatchingSkills matchingSkills = new MatchingSkills();
matchingSkills.addEvent(event); //自定义事件
CommonEventSubscribeInfo subscribeInfo = new CommonEventSubscribeInfo(matchingSkills);
```

```
subscribeInfo.setPriority(100); //设置优先级取值范围[-1000,1000],值默认为 0
MyCommonEventSubscriber subscriber = new MyCommonEventSubscriber(subscribeInfo);
try {
    CommonEventManager.subscribeCommonEvent(subscriber);
} catch (RemoteException e) {
    HiLog.error(LABEL, "Exception occurred during subscribeCommonEvent invocation.");
}
```

（3）针对在 onReceiveEvent 中不能执行耗时操作的限制，可以使用 CommonEventSubscriber 的 goAsyncCommonEvent()实现异步操作，函数返回后仍保持该公共事件活跃，且执行完成后必须调用 AsyncCommonEventResult.finishCommonEvent()结束，相关代码如下。

```
EventRunner runner = EventRunner.create();            //EventRunner 创建新线程,将耗时的操
                                                      //作放到新的线程上执行
MyEventHandler myHandler = new MyEventHandler(runner); //MyEventHandler 为 EventHandler 的派
                                                      //生类,在不同线程间分发、处理事件和
                                                      //Runnable 任务
@Override
public void onReceiveEvent(CommonEventData commonEventData){
    final AsyncCommonEventResult result = goAsyncCommonEvent();
    Runnable task = new Runnable() {
        @Override
        public void run() {
            ........                                   //待执行的操作,由开发者定义
            result.finishCommonEvent();                //调用 finish 结束异步操作
        }
    };
    myHandler.postTask(task);
}
```

4. 退订公共事件

在 Ability 的 onStop()中调用 CommonEventManager.unsubscribeCommonEvent()方法退订公共事件。调用后，之前订阅的所有公共事件均被退订，相关代码如下。

```
try {
    CommonEventManager.unsubscribeCommonEvent(subscriber);
} catch (RemoteException e) {
    HiLog.error(LABEL, "Exception occurred during unsubscribeCommonEvent invocation.");
}
```

针对公共事件开发，CommonEvent 演示了公共事件的订阅、发布和退订，示例工程参考地址为：https://gitee.com/openharmony/app_samples/tree/master/ability/CommonEvent。

3.3.2 通知开发

HarmonyOS 提供通知功能，即在一个应用的 UI 之外显示的消息，主要是提醒用户有

来自该应用中的信息。当应用向系统发出通知时,它将先以图标的形式显示在通知栏中,用户可以下拉通知栏查看详细信息。

常见的使用场景:显示接收到短消息、即时消息等;显示应用的推送消息,例如广告、版本更新等;显示当前正在进行的事件,例如播放音乐、导航、下载等。

1. 通知开发接口关系

通知相关基础类包含 NotificationSlot、NotificationRequest 和 NotificationHelper,如图 3-9 所示。

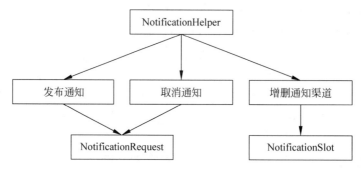

图 3-9 通知基础类关系图

1) NotificationSlot

NotificationSlot 可以对提示音、振动、重要级别等进行设置。一个应用可以创建一个或多个 NotificationSlot,在发布通知时,通过绑定不同的 NotificationSlot,实现不同用途。

NotificationSlot 需要先通过 NotificationHelper 的 addNotificationSlot(NotificationSlot)方法发布后,通知才能绑定使用;所有绑定 NotificationSlot 的通知在发布后都具备相应的特性,对象在创建后,将无法更改这些设置,对于是否启动相应设置,用户有最终控制权。

不指定 NotificationSlot 时,当前通知会使用默认的 NotificationSlot,默认的 NotificationSlot 优先级为 LEVEL_DEFAULT,主要接口及功能描述如表 3-8 所示。

表 3-8 NotificationSlot 的主要接口及功能描述

主要接口	功能描述
NotificationSlot(String id,String name,int level)	构造 NotificationSlot
setLevel(int level)	设置 NotificationSlot 的级别
setName(String name)	设置 NotificationSlot 的命名
setDescription(String description)	设置 NotificationSlot 的描述信息
enableBypassDnd(boolean bypassDnd)	设置是否绕过系统的免打扰模式
setEnableVibration(boolean vibration)	设置收到通知时是否使能振动
setEnableLight(boolean isLightEnabled)	设置收到通知时是否开启呼吸灯,前提是当前硬件支持呼吸灯
setLedLightColor(int color)	设置收到通知时的呼吸灯颜色
setSlotGroup(String groupId)	绑定当前 NotificationSlot 到一个 NotificationSlot 组

NotificationSlot 的级别由低到高支持如下几种：LEVEL_NONE 表示通知不发布；LEVEL_MIN 表示通知可以发布，但不在状态栏显示，不自动弹出，无提示音，该级别不适用于前台服务的场景；LEVEL_LOW 表示通知发布后在状态栏显示，不自动弹出，无提示音；LEVEL_DEFAULT 表示通知发布后在状态栏显示，不自动弹出，触发提示音；LEVEL_HIGH 表示通知发布后在状态栏显示，自动弹出，触发提示音。

2) NotificationRequest

NotificationRequest 用于设置具体的通知对象，包括设置通知的属性，例如通知的分发时间、小图标、大图标、自动删除等参数，以及设置具体的通知类型，例如普通文本、长文本等，主要接口及功能描述如表 3-9 所示。

表 3-9　NotificationRequest 的主要接口及功能描述

主要接口	功能描述
NotificationRequest()	构建一个通知
NotificationRequest(int notificationId)	构建一个通知，指定通知的 ID。通知的 ID 在应用内具有唯一性，如果不指定，则默认为 0
setNotificationId(int notificationId)	设置当前通知 ID
setAutoDeletedTime(long time)	设置通知自动取消的时间戳
setContent(NotificationRequest.NotificationContent content)	设置通知的具体内容
setDeliveryTime(long deliveryTime)	设置通知分发的时间戳
setSlotId(String slotId)	设置通知的 NotificationSlot ID
setTapDismissed(boolean tapDismissed)	设置通知在用户单击后是否自动取消
setLittleIcon(PixelMap smallIcon)	设置通知的小图标，在通知左上角显示
setBigIcon(PixelMap bigIcon)	设置通知的大图标，在通知的右侧显示
setGroupValue(String groupValue)	设置分组通知，相同分组的通知在通知栏显示时，将会折叠在一组应用中显示
addActionButton(NotificationActionButton actionButton)	设置通知添加 ActionButton
setIntentAgent(IntentAgent agent)	设置通知承载指定的 IntentAgent，在通知中实现即将触发的事件

目前支持 6 种通知类型：普通文本 NotificationNormalContent、长文本 NotificationLongTextContent、图片 NotificationPictureContent、多行 NotificationMultiLineContent、社交 NotificationConversationalContent、媒体 NotificationMediaContent，主要接口及功能描述如表 3-10 所示。

表 3-10　通知类型的主要接口及功能描述

类名	主要接口	功能描述
NotificationNormalContent	setTitle(String title)	设置通知标题
	setText(String text)	设置通知内容
	setAdditionalText(String additionalText)	设置通知次要内容，是对通知内容的补充

续表

类　名	主要接口	功能描述
NotificationNormalContent	setBriefText(String briefText)	设置通知概要内容，是对通知内容的总结
	setExpandedTitle(String expandedTitle)	设置附加图片通知展开时的标题
	setBigPicture(PixelMap bigPicture)	设置通知的图片内容，附加在 setText(String text)下方
NotificationLongTextContent	setLongText(String longText)	设置通知的长文本
NotificationConversationalContent	setConversationTitle(String conversationTitle)	设置社交通知的标题
	addConversationalMessage(ConversationalMessage message)	通知添加一条消息
NotificationMultiLineContent	addSingleLine(String line)	在当前通知中添加一行文本
NotificationMediaContent	setAVToken(AVToken avToken)	将媒体通知绑定指定的 AVToken
	setShownActions(int[] actions)	设置媒体通知待展示的按钮

通知发布后设置不可修改，如果下次发布通知使用相同的 ID，则会更新之前发布的通知。

3) NotificationHelper

NotificationHelper 封装了发布、更新、删除通知等静态方法，主要接口及功能描述如表 3-11 所示。

表 3-11　NotificationHelper 的主要接口及功能描述

主要接口	功能描述
publishNotification(NotificationRequest request)	发布一条通知
publishNotification(String tag, NotificationRequest request)	发布一条带 TAG 的通知
cancelNotification(int notificationId)	取消指定通知
cancelNotification(String tag, int notificationId)	取消指定带 TAG 的通知
cancelAllNotifications()	取消之前发布的所有通知
addNotificationSlot(NotificationSlot slot)	创建一个 NotificationSlot
getNotificationSlot(String slotId)	获取 NotificationSlot
removeNotificationSlot(String slotId)	删除一个 NotificationSlot
getActiveNotifications()	获取当前应用发的活跃通知
getActiveNotificationNums()	获取系统中当前应用发的活跃通知数量
setNotificationBadgeNum(int num)	设置通知的角标
setNotificationBadgeNum()	设置当前应用中活跃状态通知的数量在角标显示

2. 通知开发步骤

通知的开发分为创建 NotificationSlot、发布通知和取消通知等开发场景。

1）创建 NotificationSlot

NotificationSlot 可以设置公共通知的振动、重要级别等，并通过调用 NotificationHelper.addNotificationSlot() 发布 NotificationSlot 对象，相关代码如下。

```
NotificationSlot slot = new NotificationSlot("slot_001", "slot_default", NotificationSlot.
LEVEL_MIN); //创建 notificationSlot 对象
slot.setDescription("NotificationSlotDescription");
slot.setEnableVibration(true);                    //设置振动提醒
slot.setEnableLight(true);                        //设置开启呼吸灯提醒
slot.setLedLightColor(Color.RED.getValue());      //设置呼吸灯的提醒颜色
try {
    NotificationHelper.addNotificationSlot(slot);
} catch (RemoteException ex) {
    HiLog.error(LABEL, "Exception occurred during addNotificationSlot invocation.");
}
```

2）发布通知

发布通知步骤如下。

（1）构建 NotificationRequest 对象，应用发布通知前，通过 NotificationRequest 的 setSlotId() 方法与 NotificationSlot 绑定，使通知在发布后具备该对象的特征，相关代码如下。

```
int notificationId = 1;
NotificationRequest request = new NotificationRequest(notificationId);
request.setSlotId(slot.getId());
```

（2）调用 setContent() 设置通知的内容，相关代码如下。

```
String title = "title";
String text = "There is a normal notification content.";
NotificationNormalContent content = new NotificationNormalContent();
content.setTitle(title)
       .setText(text);
NotificationRequest.NotificationContent notificationContent = new NotificationRequest.
NotificationContent(content);
request.setContent(notificationContent); //设置通知的内容
```

（3）调用 publishNotification() 发布通知，相关代码如下。

```
try {
    NotificationHelper.publishNotification(request);
} catch (RemoteException ex) {
    HiLog.error(LABEL, "Exception occurred during publishNotification invocation.");
}
```

3）取消通知

取消通知分为取消指定单条通知和取消所有通知，应用只能取消自己发布的通知。

（1）调用 cancelNotification()取消指定的单条通知,相关代码如下。

```
int notificationId = 1;
try {
    NotificationHelper.cancelNotification(notificationId);
} catch (RemoteException ex) {
    HiLog.error(LABEL, "Exception occurred during cancelNotification invocation.");
}
```

（2）调用 cancelAllNotifications()取消所有通知,相关代码如下。

```
try {
    NotificationHelper.cancelAllNotifications();
} catch (RemoteException ex) {
    HiLog.error(LABEL, "Exception occurred during cancelAllNotifications invocation.");
}
```

针对通知开发和通知功能,在一个应用的 UI 之外显示的消息,主要是提醒用户有来自该应用中的信息,演示如何发布通知和取消通知。示例工程参考地址为：https://gitee.com/openharmony/app_samples/tree/master/ability/Notification。

3.3.3 IntentAgent 开发

IntentAgent 封装了一个指定行为的 Intent,可以通过 triggerIntentAgent 接口主动触发,也可以与通知绑定被动触发。具体行为包括启动 Ability 和发布公共事件。例如,在单击通知后跳转到一个新的 Ability,不单击则不触发。

1. IntentAgent 接口关系

IntentAgent 相关基础类包括 IntentAgentHelper、IntentAgentInfo、IntentAgentConstant 和 TriggerInfo,如图 3-10 所示。

图 3-10 IntentAgent 基础类关系图

1) IntentAgentHelper

IntentAgentHelper 封装了获取、激发、取消 IntentAgent 等静态方法,主要接口及功能描述如表 3-12 所示。

表 3-12　IntentAgentHelper 的主要接口及功能描述

主 要 接 口	功 能 描 述
getIntentAgent(Context context, IntentAgentInfo paramsInfo)	获取一个 IntentAgent 实例
triggerIntentAgent (Context context, IntentAgent agent, IntentAgent.Oncompleted onCompleted, EventHandler handler, TriggerInfo paramsInfo)	主动激发一个 IntentAgent 实例
cancel(IntentAgent agent)	取消一个 IntentAgent 实例
judgeEquality(IntentAgent agent, IntentAgent otherAgent)	判断两个 IntentAgent 实例是否相等
getHashCode(IntentAgent agent)	获取一个 IntentAgent 实例的哈希码
getBundleName(IntentAgent agent)	获取一个 IntentAgent 实例的包名
getUid(IntentAgent agent)	获取一个 IntentAgent 实例的用户 ID

2) IntentAgentInfo

IntentAgentInfo 类封装获取一个 IntentAgent 实例所需的数据。使用构造函数 IntentAgentInfo(int requestCode, OperationType operationType, List<Flags> flags, List<Intent> intents, IntentParams extraInfo) 获取 IntentAgentInfo 对象。

requestCode：使用者定义的一个私有值。

operationType：IntentAgentConstant.OperationType 枚举中的值。

flags：IntentAgentConstant.Flags 枚举中的值。

intents：将被执行的意图列表。operationType 的值为 START_ABILITY、START_SERVICE 和 SEND_COMMON_EVENT 时，intents 列表只允许包含一个 Intent；operationType 的值为 START_ABILITIES 时，intents 列表允许包含多个 Intent。

extraInfo：表明如何启动一个有页面的 ability，可以为 null，只在 operationType 的值为 START_ABILITY 和 START_ABILITIES 时有意义。

3) IntentAgentConstant

IntentAgentConstant 类中包含 OperationType 和 Flags 两个枚举类。

IntentAgentConstant.OperationType 类的枚举值如下。

(1) UNKNOWN_TYPE：不识别的类型。

(2) START_ABILITY：开启一个有页面的 Ability。

(3) START_ABILITIES：开启多个有页面的 Ability。

(4) START_SERVICE：开启一个无页面的 Ability。

(5) SEND_COMMON_EVENT：发送一个公共事件。

IntentAgentConstant.Flags 类的枚举值如下。

(1) ONE_TIME_FLAG：IntentAgent 仅能使用一次，只在 operationType 的值为 START_ABILITY、START_SERVICE 和 SEND_COMMON_EVENT 时有意义。

(2) NO_BUILD_FLAG：如果描述 IntentAgent 对象不存在，则不创建它，直接返回 null，只在 operationType 的值为 START_ABILITY、START_SERVICE 和 SEND_COMMON_EVENT

时有意义。

（3）CANCEL_PRESENT_FLAG：在生成一个新的 IntentAgent 对象前取消已存在的一个 IntentAgent 对象，只在 operationType 的值为 START_ABILITY、START_SERVICE 和 SEND_COMMON_EVENT 时有意义。

（4）UPDATE_PRESENT_FLAG：使用新的 IntentAgent 额外数据替换已存在 IntentAgent 中的额外数据，只在 operationType 的值为 START_ABILITY、START_SERVICE 和 SEND_COMMON_EVENT 时有意义。

（5）CONSTANT_FLAG：IntentAgent 是不可变的。

（6）REPLACE_ELEMENT：当前 Intent 中的 element 属性可被 IntentAgentHelper.triggerIntentAgent() 中 Intent 的 element 属性取代。

（7）REPLACE_ACTION：当前 Intent 中的 Action 属性可被 IntentAgentHelper.triggerIntentAgent() 中 Intent 的 Action 属性取代。

（8）REPLACE_URI：当前 Intent 中的 URI 属性可被 IntentAgentHelper.triggerIntentAgent() 中 Intent 的 URI 属性取代。

（9）REPLACE_ENTITIES：当前 Intent 中的 entities 属性可被 IntentAgentHelper.triggerIntentAgent() 中 Intent 的 entities 属性取代。

（10）REPLACE_BUNDLE：当前 Intent 中的 bundleName 属性可被 IntentAgentHelper.triggerIntentAgent() 中 Intent 的 bundleName 属性取代。

4）TriggerInfo

TriggerInfo 类封装了主动激发一个 IntentAgent 实例所需的数据，使用构造函数 TriggerInfo(String permission、IntentParams extraInfo、Intent intent、int code) 获取 TriggerInfo 对象。

（1）String permission：IntentAgent 的接收者权限名称，只在 operationType 的值为 SEND_COMMON_EVENT 时，参数才有意义。

（2）IntentParams extraInfo：激发 IntentAgent 时用户自定义的额外数据。

（3）Intent intent：额外的 Intent。如果 IntentAgentInfo 成员变量 flags 包含 CONSTANT_FLAG，则忽略该参数；如果 flags 包含 REPLACE_ELEMENT、REPLACE_ACTION、REPLACE_URI、REPLACE_ENTITIES 或 REPLACE_BUNDLE，则使用额外 Intent 的 element、action、uri、entities 或 bundleName 属性替换原始 Intent 中对应的属性。如果 Intent 为空，则不替换原始 Intent 的属性。

（4）int code：提供给 IntentAgent 目标的结果码。

2. IntentAgent 开发步骤

IntentAgent 开发步骤如下。

1）获取 IntentAgent 代码

```
//指定要启动的 Ability 的 BundleName 和 AbilityName 字段
//将 Operation 对象设置到 Intent 中
```

```
Operation operation = new Intent.OperationBuilder()
        .withDeviceId("")
        .withBundleName("com.testintentagent")
        .withAbilityName("com.testintentagent.entry.IntentAgentAbility")
        .build();
intent.setOperation(operation);
List < Intent > intentList = new ArrayList <>();
intentList.add(intent);
//定义请求码
int requestCode = 200;
//设置 flags
List < IntentAgentConstant.Flags > flags = new ArrayList <>();
flags.add(IntentAgentConstant.Flags.UPDATE_PRESENT_FLAG);
//指定启动一个有页面的 Ability
IntentAgentInfo paramsInfo = new IntentAgentInfo(requestCode, IntentAgentConstant.OperationType.
START_ABILITY, flags, intentList, null);
//获取 IntentAgent 实例
IntentAgent agent = IntentAgentHelper.getIntentAgent(this, paramsInfo);
```

2)通知中添加 IntentAgent 的代码

```
int notificationId = 1;
NotificationRequest request = new NotificationRequest(notificationId);
String title = "title";
String text = "There is a normal notification content.";
NotificationRequest.NotificationNormalContent content = new NotificationRequest.
NotificationNormalContent();
content.setTitle(title)
        .setText(text);
NotificationContent notificationContent = new NotificationContent(content);
request.setContent(notificationContent);       //设置通知的内容
request.setIntentAgent(agent);                  //设置通知的 IntentAgent
```

3)主动激发 IntentAgent 的代码

```
int code = 100;
IntentAgentHelper.triggerIntentAgent(this, agent, null, null, new TriggerInfo(null, null,
null, code));
```

针对 IntentAgent 开发指导,演示如何通过 IntentAgent 启动 Ability 和发布公共事件。示例工程参考地址为:https://gitee.com/openharmony/app_samples/tree/master/ability/IntentAgent。

3.3.4 后台代理定时提醒开发

在应用开发时,可以调用后台代理提醒类 ReminderRequest 创建定时提醒,包括倒计时、日历和闹钟 3 种类型。使用后台代理提醒功能后,应用可以被冻结或退出,计时和弹出提醒的功能将被后台系统服务代理。

1. 后台代理定时提醒接口关系

ReminderRequest 涉及的基础类包括 ReminderHelper、ReminderRequestTimer、ReminderRequestCalendar 和 ReminderRequestAlarm，如图 3-11 所示。

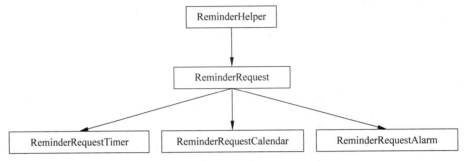

图 3-11　ReminderRequest 基础类关系图

1）ReminderHelper

ReminderHelper 是封装发布、取消提醒类通知的方法，其主要接口及功能描述如表 3-13 所示。

表 3-13　ReminderHelper 的主要接口及功能描述

主要接口	功能描述
public static int publishReminder(ReminderRequest reminderReq) throws RemoteException, ReminderManager.AppLimitExceedsException, ReminderManager.SysLimitExceedsException	发布一个定时提醒类通知 ReminderManager.AppLimitExceedsException 系统中保存的当前应用有效的提醒个数超出最大限制数量 30 个时抛出（不包括已经超时，即后续不会再提醒的实例） ReminderManager.SysLimitExceedsException 中保存的整个系统有效的提醒个数超出最大限制数量 2000 个时抛出（不包括已经超时，即后续不再提醒的实例）
public static void addNotificationSlot(NotificationSlot slot) throws RemoteException	注册一个提醒类需要使用的 NotificationSlot
public static void cancelReminder(int reminderId) throws RemoteException	取消一个指定的提醒类通知（reminderId 从 publishReminder 的返回值获取）
public static void removeNotificationSlot(String slotId) throws RemoteException	删除一个 slot 实例
public static List<ReminderRequest> getValidReminders() throws RemoteException	获取当前应用设置的所有有效提醒
public static void cancelAllReminders() throws RemoteException	取消当前应用设置的所有提醒

2）ReminderRequest

ReminderRequest 是后台代理提醒类基类，封装提醒相关的属性查询和设置的操作，主要接口及功能描述如表 3-14 所示。

表 3-14 ReminderRequest 的主要接口及功能描述

主 要 接 口	功 能 描 述
public long getRingDuration()	获取设置的提醒时长，单位为 s，例如设置开始响铃后的时长
public int getSnoozeTimes()	获取设置的延迟提醒次数
public long getTimeInterval()	获取设置的延迟提醒间隔
public ReminderRequest setRingDuration (long ringDurationInSeconds)	设置提醒时长，单位为 s，例如设置开始响铃后的时长
public ReminderRequest setSnoozeTimes (int snoozeTimes)	设置延迟提醒的次数（倒计时设置延迟提醒无效）
public ReminderRequest setTimeInterval (long timeIntervalInSeconds)	设置延迟提醒的时间间隔（倒计时设置延迟提醒无效）
public ReminderRequest setActionButton (String title, int type)	在提醒弹出的通知界面中添加 NotificationActionButton
public ReminderRequest setIntentAgent (String pkgName, String abilityName)	设置单击通知信息后需要跳转目标包的信息
public ReminderRequest setMaxScreenIntentAgent (String pkgName, String abilityName)	设置提醒到达时跳转的目标包。如果设备正在使用中，则弹出一个通知框
public String getTitle()	获取提醒的标题
public ReminderRequest setTitle(String title)	设置提醒的标题
public String getContent()	获取提醒的内容
public ReminderRequest setContent(String content)	设置提醒的内容
public String getExpiredContent()	获取提醒"过期"时显示的扩展内容
public ReminderRequest setExpiredContent (String expiredContent)	设置提醒"过期"时显示的扩展内容
public String getSnoozeContent()	获取提醒"再响"时显示的扩展内容
public ReminderRequest setSnoozeContent (String snoozeContent)	设置提醒"再响"时显示的扩展内容
public int getNotificationId()	获取提醒使用 notificationRequest 的 ID，参见 NotificationRequest.setNotificationId(int id)
public ReminderRequest setNotificationId (int notificationId)	设置提醒使用 notificationRequest 的 ID
public String getSlotId()	获取提醒使用的 slot ID
public String SetSlotId(String slotId)	设置提醒使用的 slot ID

3) ReminderRequestTimer

ReminderRequestTimer 为提醒类子类,用于倒计时提醒,主要接口及功能描述如表 3-15 所示。

表 3-15　ReminderRequestTimer 的主要接口及功能描述

主要接口	功能描述
public ReminderRequestTimer(long triggerTimeInSeconds)	创建一个倒计时提醒实例,经过指定时间后触发提醒

4) ReminderRequestCalendar

ReminderRequestCalendar 为提醒类子类,用于日历类提醒。可以指定提醒时间精确为某年某月某日某时某分,也可以指定哪些月份的哪些天的同一时间重复提醒,主要接口及功能描述如表 3-16 所示。

表 3-16　ReminderRequestCalendar 的主要接口及功能描述

主要接口	功能描述
public ReminderRequestCalendar(LocalDateTime dateTime,int[] repeatMonths,int[] repeatDays)	创建一个日历类提醒实例,在指定的时间触发提醒

5) ReminderRequestAlarm

ReminderRequestAlarm 为提醒类子类,用于闹钟类提醒。可以指定几点几分提醒,或者每周哪几天指定时间提醒。主要接口及功能描述如表 3-17 所示。

表 3-17　ReminderRequestAlarm 的主要接口及功能描述

主要接口	功能描述
public ReminderRequestAlarm(int hour, int minute,int[] daysOfWeek)	创建一个闹钟类提醒实例,在指定的时间触发提醒

2. 后台代理定时提醒开发步骤

开发步骤如下。

1) 声明使用权限

使用后台代理提醒需要在配置文件中声明需要此权限。

```
"reqPermissions":[{"name":"ohos.permission.PUBLISH_AGENT_REMINDER"}]
```

2) 创建提醒步骤

创建提醒共 8 个步骤,具体方法如下。

(1) 设置渠道信息。

```
NotificationSlot slot = new NotificationSlot("slot_id", "slot_name", NotificationSlot.LEVEL_HIGH);
```

```
slot.setEnableLight(false);
slot.setEnableVibration(true);
```

(2) 向代理服务添加渠道对象。

```
try {
    ReminderHelper.addNotificationSlot(slot);
} catch (RemoteException e) {
    e.printStackTrace();
}
```

(3) 创建提醒类通知对象。

```
int[] repeatDay = {};
ReminderRequest reminder = new ReminderRequestAlarm(10, 30, repeatDay);
```

(4) 设置提醒内容。

```
reminder.setTitle("set title here").setContent("set content here");
```

(5) 设置提醒时长等属性。

```
reminder.setSnoozeTimes(1).setTimeInterval(5 * 60).setRingDuration(10);
```

(6) 设置 IntentAgent（假设包名为 com.ohos.aaa，Ability 类名为 FirstAbility）。

```
reminder.setIntentAgent("com.ohos.aaa", FirstAbility.class.getName());
```

(7) 设置提醒信息框中的"延迟提醒"和"关闭"按钮（可选）（ActionButton）。

```
reminder.setActionButton("snooze", ReminderRequest.ACTION_BUTTON_TYPE_SNOOZE).
    setActionButton("close", ReminderRequest.ACTION_BUTTON_TYPE_CLOSE);
```

(8) 发布提醒类通知。

```
try {
    ReminderHelper.publishReminder(reminder);
} catch (ReminderManager.AppLimitExceedsException e) {
    e.printStackTrace();
} catch (ReminderManager.SysLimitExceedsException e) {
    e.printStackTrace();
} catch (RemoteException e) {
    e.printStackTrace();
}
```

3）创建倒计时提醒示例

```
//经过 1 分钟后提醒
ReminderRequest reminderRequestTimer = new ReminderRequestTimer(60);
```

4）创建一次性日历提醒

```
//2021 年 3 月 2 日 14 点 30 分提醒
int[] repeatMonths = {};
int[] repeatDays = {};
ReminderRequestCalendar reminderRequestCalendar = new ReminderRequestCalendar( LocalDateTime. of
(2021, 3, 2, 14, 30), repeatMonths, repeatDays);
```

5）创建重复日历提醒

```
// 3 月份、5 月份的 9 号和 15 号 14 点 30 分提醒,延迟 10 分钟后再次提醒,默认延迟次数为 3 次
int[] repeatMonths = {3, 5};
int[] repeatDaysOfMonth = {9, 15};
ReminderRequestCalendar reminderRequestCalender = new ReminderRequestCalendar( LocalDateTime. of
(2021, 3, 2, 14, 30), repeatMonths, repeatDaysOfMonth);
reminderRequestCalender.setTimeInterval(10 * 60);
```

6）创建一次性闹钟提醒

```
// 13 点 59 分提醒,如果当前时间大于 13 点 59 分,则取后一天的 13 点 59 分
int[] repeatDay = {};
ReminderRequest reminderRequestAlarm = new ReminderRequestAlarm(13, 59, repeatDay);
```

7）创建重复的闹钟提醒

```
//每周 1、2、3、4 的 13 点 59 分提醒
int[] repeatDay = {1, 2, 3, 4};
ReminderRequest reminderRequestAlarm = new ReminderRequestAlarm(13, 59, repeatDay);
```

8）创建用于延迟提醒的 ActionButton 界面

```
reminderRequest.setActionButton("snooze", ReminderRequest.ACTION_BUTTON_TYPE_SNOOZE);
```

9）创建用于关闭提醒的 ActionButton 界面

```
reminderRequest.setActionButton("close", ReminderRequest.ACTION_BUTTON_TYPE_CLOSE);
```

notificationId 相同的不同通知请求,在通知栏展示的内容会被覆盖,对于提醒来说,可能不希望被覆盖,开发时可以注意设置不同的 notificationId。倒计时不支持持久化,系统重启后,所有倒计时失效。

3.4 后台任务调度和管控

对于有用户交互的 OS 来说,资源优先分配给与用户交互的业务进程,换句话说,在支撑 OS 运行的进程以外,用户能感知到的业务进程优先级最高,所以后台任务调度控制的范围是用户感知不到的业务进程。

HarmonyOS 将应用的资源使用生命周期划分为前台、后台和挂起 3 个阶段。前台运

行不受资源调度的约束，后台会根据应用业务的具体任务情况进行资源使用管理，在挂起状态时，会对应用的资源使用进行调度和控制约束，以保障其他体验类业务对资源的竞争使用。

后台任务调度和管控主要是对在后台状态下的资源使用进行控制，应用从前台退到后台，可能有各种业务需求，为了达到系统资源使用能效最优的目的，HarmonyOS 提供了后台任务能力。

后台任务特指应用或业务模块处于后台（无可见界面）时，需要继续执行或者后续执行的业务。为避免后台任务调度和管控对业务执行的影响，HarmonyOS 将后台任务分为 3 种类型。

无后台业务：退出后台后，无任务需要处理。

短时任务：退出后台后，如果有紧急不可推迟且短时间能完成的任务，应用退出后台要进行数据压缩，不可中断，则使用短时任务申请延迟进入挂起（Suspend）状态。

长驻任务：如果用户发起的可感知业务需要长时间后台运行，例如后台播放音乐、导航、上传下载、设备连接、VoIP 等，则使用长驻任务避免进入挂起状态。

3.4.1 短时任务

退到后台的应用有不可中断且短时间能完成的任务时，可以使用短时任务机制，该机制允许应用在后台短时间内完成任务，保障应用业务运行不受后台生命周期管理的影响。

短时任务仅针对应用的临时任务提供资源使用生命周期保障，限制单次最大使用时长为 3min，全天使用配额默认为 10min（具体时长系统根据应用场景和系统状态智能调整）。接口说明参考地址为：https://developer.harmonyos.com/cn/docs/documentation/doc-references/backgroundtaskmanager-0000001054440069。

短时任务的使用需要遵从如下约束和规则。

申请时机：允许应用在前台或退后台在被挂起之前（应用退到后台默认有 6~12s 的运行时长，具体时长由系统根据具体场景决定）申请延迟挂起，否则可能被挂起，导致申请失败。

超时：延迟挂起超时（Timeout），系统通过回调通知应用，应用需要取消对应的延迟挂起，或再次申请延迟挂起。超期不取消或不处理，该应用会被强制取消延迟挂起。

取消时机：任务完成后申请方应用主动取消延时申请，不要等到超时后被系统取消，否则会影响该应用的后台允许运行时长配额。

配额机制：为了防止应用滥用保活，或者申请后不取消，每个应用每天都会有一定配额（根据用户的使用习惯动态调整），配额消耗完不再允许申请短时任务，所以应用完成短时任务后立刻取消延时申请，避免消耗配额（此配额是指申请的时长，系统默认应用在后台运行的时间不计算在内）。

3.4.2 长驻任务

长驻任务类型给用户能直观感知到的且需要一直在后台运行的业务提供后台运行生命周期的保障。例如，业务需要在后台播放声音或者持续导航定位等，此类用户能够感知到后台业务行为，可以通过使用长驻任务对应的后台模式保障业务在后台运行，支撑应用完成在后台的业务。

1. 后台模式分类

HarmonyOS 提供 10 种后台模式，供需要在后台做长驻任务的业务使用，如表 3-18 所示。

表 3-18 长驻任务后台模式及功能描述

长驻任务后台模式	英 文 名	功 能 描 述
数据传输	data-transfer	通过网络/对端设备进行数据下载、备份、分享、传输等业务
播音	audio-playback	音频输出业务
录音	audio-recording	音频输入业务
画中画	picture-in-picture	画中画、小窗口播放视频业务
音视频通话	voip	音视频电话、VoIP 业务
导航/位置更新	location	定位、导航业务
蓝牙设备连接及传输	bluetooth-interaction	蓝牙扫描、连接、传输业务
WLAN 设备连接及传输	wifi-interaction	WLAN 扫描、连接、传输业务
屏幕抓取	screen-fetch	录屏、截屏业务
多设备互联	multiDeviceConnection	多设备互联、分布式调度和迁移等

2. 使用长驻任务

HarmonyOS 应用开发工具 DevEco Studio 在业务 ServiceAbility 创建时提供后台模式的选择，针对当前创建的 ServiceAbility 可以赋予对应的后台模式类型设置，如图 3-12 所示。

根据业务需要选择对应的后台模式后，会在应用的 config.json 文件中新创建的 ServiceAbility 组件下生成对应选择的后台模式配置，如图 3-13 所示，只有 ServiceAbility 才有对应的后台模式类型选择和配置。

在 Service 创建的方法中调用 keepBackgroundRunning()，将 Service 与通知绑定。

调用 keepBackgroundRunning() 方法前需要在配置文件中声明 ohos.permission.KEEP_BACKGROUND_RUNNING 权限。完成对应的后台业务后，在销毁服务的方法中调用 cancelBackgroundRunning() 方法，即可停止使用长驻任务。

3. 长驻任务使用约束

如果用户选择可感知业务，例如播音、导航、上传、下载等，触发对应后台模式，在任务启动或退入后台时，需要提醒用户。

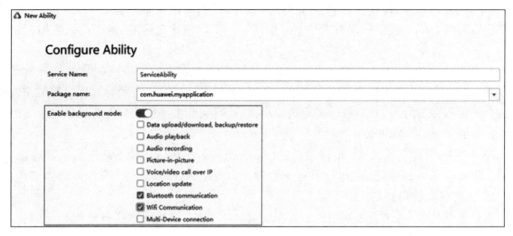

图 3-12　后台模式类型设置

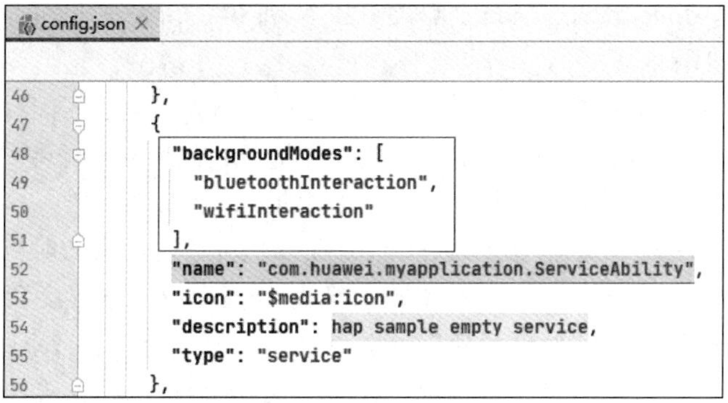

图 3-13　后台模式类型选择和配置

如果任务结束,应用会主动退出后台模式。若在后台运行期间,系统监测到应用并未使用对应后台模式的资源,则会被挂起。

避免不合理地申请后台长驻任务,长驻任务类型要与应用的类型匹配,如果执行的任务和申请的类型不匹配,也会被系统检测到并被挂起。

长驻任务是为了真正在后台长时间执行某任务,如果一个应用申请了长驻任务,但在实际运行过程中,并未真正运行或执行此类任务,也会被系统检测到并被挂起。

3.4.3　托管任务

托管任务是系统提供的一种后台代理机制。通过系统提供的代理 API 接口,用户可以把任务交由系统托管,例如后台下载、定时提醒、后台非持续定位。

1. 托管任务类型

托管任务-后台非持续定位(non-sustained Location):如果应用未申请 location 长驻模

式,且在后台依然尝试获取位置信息,此时应用行为被视为使用非持续定位能力,后台非持续定位限制每 30min 提供一次位置信息。应用不需要高频次定位时,建议优先使用非持续定位。

托管任务-后台提醒代理(Reminder):后台提醒代理主要提供了一种机制,使开发者在应用开发时,可以调用这些接口去创建定时提醒,包括倒计时、日历、闹钟三种提醒类型。使用后台代理提醒能力后,应用可以被冻结或退出,计时和弹出提醒的功能将被后台系统服务代理。

托管任务-后台下载代理:系统提供 DownloadSession 接口实现下载任务代理功能,应用提交下载任务后,应用被退出,下载任务仍然可以继续执行,且支持下载任务断点续传。

2. 托管任务使用约束

后台下载代理,系统会根据用户场景和设备状态,对不同的下载任务进行相应的管控,避免影响功耗和性能。

后台提醒需要申请 ohos. permission. PUBLISH_AGENT_REMINDER 权限,后台非持续定位需要申请 ohos. permission. LOCATION 和 ohos. permission. LOCATION_IN_BACKGROUND 权限。

资源滥用会影响系统性能和功耗,托管任务类型要与应用类型匹配,按照华为应用市场分类方法匹配规则,分为游戏类和应用类。游戏类没有后台长驻任务申请标准和后台托管任务申请标准,应用类的不同子类具有不同的限制约束。

原则上所有应用都会受后台任务调度和管控的约束,只是约束的时机和状态不同而已。系统提供了选择设置的方式供用户从使用需求维度干预后台任务调度和管控,以华为手机为例设置路径:手机管家→应用启动管理→选择手动管理→允许后台活动。

应用后台任务调度和管控以支撑用户使用体验为第一优先级,结合用户使用习惯、用户使用情景状态、设备模式状态、应用分类、应用资源占用统计等多个维度综合判断应用后台任务调度和管控的优先级。后台任务调度和管控机制仅对 HarmonyOS 应用进行调度和管控。

3.5 线程管理开发

不同应用在各自独立的进程中运行。当应用以任意形式启动时,系统为其创建进程,该进程将持续运行。当进程完成当前任务处于等待状态,且系统资源不足时,系统自动回收。

在启动应用时,系统会为该应用创建一个称为"主线程"的执行线程。该线程随着应用创建或消失,是应用的核心线程。UI 的显示和更新等操作,都在主线程上进行。主线程又称 UI 线程,默认情况下,所有的操作都在主线程上执行。如果需要执行比较耗时的任务,可创建其他线程进行处理,例如下载文件、查询数据库。

如果应用的业务逻辑比较复杂,可能需要创建多个线程执行多个任务。这种情况下,代码复杂难以维护,任务与线程的交互也会更加繁杂。要解决此问题,可以使用 TaskDispatcher 分发不同的任务。

3.5.1 线程管理开发接口关系

TaskDispatcher 是一个任务分发器,它是 Ability 分发任务的基本接口,隐藏任务所在线程的实现细节。为保证应用有更好的响应性,需要设计任务的优先级。在 UI 线程上运行的任务默认以高优先级运行,如果某个任务无须等待结果,则可以用低优先级,如表 3-19 所示。

表 3-19 线程优先级及功能描述

优 先 级	功 能 描 述
HIGH	最高任务优先级,比默认优先级、低优先级的任务有更高的概率得到执行
DEFAULT	默认任务优先级,比低优先级的任务有更高的概率得到执行
LOW	低任务优先级,比高优先级、默认优先级的任务有更低的概率得到执行

TaskDispatcher 具有多种实现,每种实现对应不同的任务分发器。在分发任务时可以指定任务的优先级,由同一任务分发器分发出的任务具有相同的优先级。系统提供的任务分发器有 GlobalTaskDispatcher、ParallelTaskDispatcher、SerialTaskDispatcher 和 SpecTaskDispatcher。

1. GlobalTaskDispatcher

全局并发任务分发器,由 Ability 执行 getGlobalTaskDispatcher() 获取。适用于任务之间没有联系的情况。一个应用只有一个 GlobalTaskDispatcher,它在程序结束时才被销毁,相关代码如下。

```
TaskDispatcher globalTaskDispatcher = getGlobalTaskDispatcher(TaskPriority.DEFAULT);
```

2. ParallelTaskDispatcher

并发任务分发器,由 Ability 执行 createParallelTaskDispatcher() 创建并返回。与 GlobalTaskDispatcher 不同的是,ParallelTaskDispatcher 不具有全局唯一性,可以创建多个。在创建或销毁 dispatcher 时,需要持有对应的对象引用,相关代码如下。

```
String dispatcherName = "parallelTaskDispatcher";
TaskDispatcher parallelTaskDispatcher = createParallelTaskDispatcher(dispatcherName, TaskPriority.DEFAULT);
```

3. SerialTaskDispatcher

串行任务分发器,由 Ability 执行 createSerialTaskDispatcher() 创建并返回。该分发器分发的所有任务都按顺序执行,但是执行这些任务的线程并不是固定的。如果要执行并行任务,应使用 ParallelTaskDispatcher 或者 GlobalTaskDispatcher,而不是创建多个 SerialTaskDispatcher。如果任务之间没有依赖,应使用 GlobalTaskDispatcher 实现。它的创建和销毁由开发者自己管理,在使用期间需要持有该对象引用,相关代码如下。

```
String dispatcherName = "serialTaskDispatcher";
TaskDispatcher serialTaskDispatcher = createSerialTaskDispatcher(dispatcherName, TaskPriority.DEFAULT);
```

4. SpecTaskDispatcher

专有任务分发器是绑定到专有线程上的任务分发器。目前已有的专有线程为 UI 线程，通过 UITaskDispatcher 进行任务分发。

UITaskDispatcher：绑定到应用主线程的专有任务分发器，由 Ability 执行 getUITaskDispatcher()创建并返回。此分发器分发的所有任务都在主线程上按顺序执行，它在应用程序结束时被销毁，相关代码如下。

```
TaskDispatcher uiTaskDispatcher = getUITaskDispatcher();
```

3.5.2 线程管理开发步骤

线程管理开发步骤如下。

1. syncDispatch

同步派发任务：派发任务并在当前线程等待任务执行完成。在返回前，当前线程会被阻塞。使用 GlobalTaskDispatcher 派发同步任务相关代码如下。

```
TaskDispatcher globalTaskDispatcher = getGlobalTaskDispatcher(TaskPriority.DEFAULT);
globalTaskDispatcher.syncDispatch(new Runnable() {
    @Override
    public void run() {
        HiLog.info(LABEL_LOG, "sync task1 run");
    }
});
HiLog.info(LABEL_LOG, "after sync task1");
globalTaskDispatcher.syncDispatch(new Runnable() {
    @Override
    public void run() {
        HiLog.info(LABEL_LOG, "sync task2 run");
    }
});
HiLog.info(LABEL_LOG, "after sync task2");
globalTaskDispatcher.syncDispatch(new Runnable() {
    @Override
    public void run() {
        HiLog.info(LABEL_LOG, "sync task3 run");
    }
});
HiLog.info(LABEL_LOG, "after sync task3");
```

执行结果如下。

```
sync task1 run
after sync task1
sync task2 run
after sync task2
sync task3 run
after sync task3
```

如果对 syncDispatch 使用不当,将会导致死锁,可能导致死锁发生的情形如下。

专有线程上,利用该分发器进行 syncDispatch。在被某个串行任务分发器(dispatcher_a)派发的任务中,再次利用同一个串行任务分发器(dispatcher_a)对象派发任务。在被某个串行任务分发器(dispatcher_a)派发的任务中,经过数次派发任务,最终又利用(dispatcher_a)串行任务分发器派发任务。例如,dispatcher_a 派发的任务使用 dispatcher_b 进行任务的派发,在 dispatcher_b 派发的任务中又利用 dispatcher_a 进行派发任务。串行任务分发器(dispatcher_a)派发的任务中利用串行任务分发器(dispatcher_b)进行同步派发任务,同时 dispatcher_b 派发的任务中利用串行任务分发器(dispatcher_a)进行同步派发任务。在特定的线程执行顺序下将导致死锁。

2. asyncDispatch

异步派发任务:派发任务并立即返回,返回值是一个可用于取消任务的接口。GlobalTaskDispatcher 派发异步任务相关代码如下。

```
TaskDispatcher globalTaskDispatcher = getGlobalTaskDispatcher(TaskPriority.DEFAULT);
Revocable revocable = globalTaskDispatcher.asyncDispatch(new Runnable() {
    @Override
    public void run() {
        HiLog.info(LABEL_LOG, "async task1 run");
    }
});
HiLog.info(LABEL_LOG, "after async task1");
```

可能的执行结果如下。

```
after async task1
async task1 run
```

3. delayDispatch

异步延迟派发任务:异步执行,函数立即返回,内部会在延时指定时间后将任务派发到相应队列中。延时时间参数仅代表在这段时间以后任务分发器会将任务加入队列中,任务的实际执行时间可能晚于这个时间。具体比这个数值晚多久,取决于队列及内部线程池的繁忙情况。GlobalTaskDispatcher 延迟派发任务相关代码如下。

```
final long callTime = System.currentTimeMillis();
final long delayTime = 50L;
TaskDispatcher globalTaskDispatcher = getGlobalTaskDispatcher(TaskPriority.DEFAULT);
Revocable revocable = globalTaskDispatcher.delayDispatch(new Runnable() {
```

```
        @Override
        public void run() {
            HiLog.info(LABEL_LOG, "delayDispatch task1 run");
            final long actualDelay = System.currentTimeMillis() - callTime;
            HiLog.info(LABEL_LOG, "actualDelayTime >= delayTime: %{public}b", (actualDelay >=
delayTime));
        }
    }, delayTime);
    HiLog.info(LABEL_LOG, "after delayDispatch task1");
```

可能的执行结果如下。

```
after delayDispatch task1
delayDispatch task1 run
actualDelayTime >= delayTime : true
```

4. Group

任务组：表示一组任务，且该组任务之间有一定的联系，由 TaskDispatcher 执行 createDispatchGroup 创建并返回。将任务加入任务组，返回一个用于取消任务的接口。将一系列相关联的下载任务放入一个任务组，执行完下载任务后关闭应用，任务组的使用方式代码如下。

```
String dispatcherName = "parallelTaskDispatcher";
TaskDispatcher dispatcher = createParallelTaskDispatcher(dispatcherName, TaskPriority.
DEFAULT);
//创建任务组
Group group = dispatcher.createDispatchGroup();
//将任务 1 加入任务组，返回一个用于取消任务的接口
dispatcher.asyncGroupDispatch(group, new Runnable() {
    @Override
    public void run() {
        HiLog.info(LABEL_LOG, "download task1 is running");
    }
});
//将与任务 1 相关联的任务 2 加入任务组
dispatcher.asyncGroupDispatch(group, new Runnable() {
    @Override
    public void run() {
        HiLog.info(LABEL_LOG, "download task2 is running");
    }
});
//在任务组中的所有任务执行完成后执行指定任务
dispatcher.groupDispatchNotify(group, new Runnable() {
    @Override
    public void run() {
        HiLog.info(LABEL_LOG, "the close task is running after all tasks in the group are
completed");
```

}
});
```

一种可能的执行结果如下。

```
download task1 is running
download task2 is running
the close task is running after all tasks in the group are completed
```

另一种可能的执行结果如下。

```
download task2 is running
download task1 is running
the close task is running after all tasks in the group are completed
```

5. Revocable

取消任务：Revocable 是取消一个异步任务的接口。异步任务包括通过 asyncDispatch、delayDispatch 和 asyncGroupDispatch 派发的任务。如果任务已经在执行中或执行完成，则会返回取消失败，相关代码如下。

```
TaskDispatcher dispatcher = getUITaskDispatcher();
Revocable revocable = dispatcher.delayDispatch(new Runnable() {
    @Override
    public void run() {
        HiLog.info(LABEL_LOG, "delay dispatch");
    }
}, 10);
boolean revoked = revocable.revoke();
HiLog.info(LABEL_LOG, " %{public}b", revoked);
```

6. syncDispatchBarrier

同步设置屏障任务：在任务组上设立任务执行屏障，同步等待任务组中的所有任务执行完成，再执行指定任务。在全局并发任务分发器（GlobalTaskDispatcher）上同步设置任务屏障，将不会起到屏障作用，同步设置屏障代码如下。

```
String dispatcherName = "parallelTaskDispatcher";
TaskDispatcher dispatcher = createParallelTaskDispatcher(dispatcherName, TaskPriority.DEFAULT);
//创建任务组
Group group = dispatcher.createDispatchGroup();
//将任务加入任务组,返回一个用于取消任务的接口
dispatcher.asyncGroupDispatch(group, new Runnable() {
    @Override
    public void run() {
        HiLog.info(LABEL_LOG, "task1 is running");   //1
    }
});
```
```

```java
dispatcher.asyncGroupDispatch(group, new Runnable() {
    @Override
    public void run() {
        HiLog.info(LABEL_LOG, "task2 is running");   //2
    }
});
dispatcher.syncDispatchBarrier(new Runnable() {
    @Override
    public void run() {
        HiLog.info(LABEL_LOG, "barrier");   //3
    }
});
HiLog.info(LABEL_LOG, "after syncDispatchBarrier");   //4
```

1 和 2 的执行顺序不定；3 和 4 总是在 1 和 2 之后按顺序执行，一种可能的执行结果如下。

```
task1 is running
task2 is running
barrier
after syncDispatchBarrier
```

另外一种执行结果如下。

```
task2 is running
task1 is running
barrier
after syncDispatchBarrier
```

7. asyncDispatchBarrier

异步设置屏障任务：在任务组上设立任务执行屏障后直接返回，指定任务将在任务组中所有任务执行完成后再执行。在全局并发任务分发器（GlobalTaskDispatcher）上异步设置任务屏障，将不会起到屏障作用。可以使用并发任务分发器（ParallelTaskDispatcher）分离不同的任务组，实现微观并行、宏观串行，异步设置屏障代码如下。

```java
TaskDispatcher dispatcher = createParallelTaskDispatcher("dispatcherName", TaskPriority.DEFAULT);
//创建任务组
Group group = dispatcher.createDispatchGroup();
//将任务加入任务组，返回一个用于取消任务的接口
dispatcher.asyncGroupDispatch(group, new Runnable() {
    @Override
    public void run() {
        HiLog.info(LABEL_LOG, "task1 is running");   //1
    }
});
dispatcher.asyncGroupDispatch(group, new Runnable() {
```

```
        @Override
        public void run() {
            HiLog.info(LABEL_LOG, "task2 is running");    //2
        }
    });
    dispatcher.asyncDispatchBarrier(new Runnable() {
        @Override
        public void run() {
            HiLog.info(LABEL_LOG, "barrier");    //3
        }
    });
    HiLog.info(LABEL_LOG, "after asyncDispatchBarrier");    //4
```

1 和 2 的执行顺序不定,但总在 3 之前执行;4 不需要等待 1、2、3 执行完成,可能的执行结果如下。

```
task1 is running
task2 is running
after asyncDispatchBarrier
barrier
```

8. applyDispatch

对指定任务执行多次的相关代码如下。

```
final int total = 10;
final CountDownLatch latch = new CountDownLatch(total);
final List<Long> indexList = new ArrayList<>(total);
TaskDispatcher dispatcher = getGlobalTaskDispatcher(TaskPriority.DEFAULT);
//执行任务 total 次
dispatcher.applyDispatch((index) -> {
    indexList.add(index);
    latch.countDown();
}, total);
//设置任务超时
try {
    latch.await();
} catch (InterruptedException exception) {
    HiLog.error(LABEL_LOG, "latch exception");
}
HiLog.info(LABEL_LOG, "list size matches, %{public}b", (total == indexList.size()));
```

执行结果如下。

```
list size matches, true
```

针对线程管理,演示如何使用 TaskDispatcher 分发不同任务,例如同步派发、异步派发、异步延迟派发等。示例工程参考地址为:https://gitee.com/openharmony/app_

samples/tree/master/thread/TaskDispatcher。

3.6 线程间通信

开发过程中，需要在当前线程中处理下载任务等较为耗时的操作，但是又不希望当前的线程受到阻塞。此时，可以使用 EventHandler 机制。EventHandler 是 HarmonyOS 用于处理线程间通信的一种机制，可以通过 EventRunner 创建新线程，将耗时的操作放到新线程上执行。这样既不阻塞原来的线程，任务又可以得到合理的处理。例如，主线程使用 EventHandler 创建子线程，子线程做耗时的下载图片操作，下载完成后，子线程通过 EventHandler 通知主线程，主线程再更新 UI。

3.6.1 概述

EventRunner 是一种事件循环器，循环处理从 EventRunner 创建新线程的事件队列中获取 InnerEvent 事件或者 Runnable 任务，InnerEvent 是 EventHandler 投递的事件。

EventHandler 是一种用户在当前线程上投递 InnerEvent 事件或者 Runnable 任务到异步线程上处理的机制。每个 EventHandler 和指定的 EventRunner 所创建的新线程绑定，并且新线程内部有一个事件队列。EventHandler 可以投递指定的 InnerEvent 事件或 Runnable 任务到这个事件队列。EventRunner 从事件队列里循环地取出事件，如果取出的是 InnerEvent 事件，将在 EventRunner 所在线程执行 processEvent 回调；如果取出的是 Runnable 任务，将在 EventRunner 所在线程执行 Runnable 的 run 回调。

EventHandler 有两个主要作用：在不同线程间分发和处理 InnerEvent 事件或 Runnable 任务；延迟处理 InnerEvent 事件或 Runnable 任务。

EventHandler 的运作机制如图 3-14 所示。使用 EventHandler 实现线程间通信的主要流程如下。

（1）EventHandler 投递具体的 InnerEvent 事件或者 Runnable 任务到 EventRunner 所创建线程的事件队列。

（2）EventRunner 循环从事件队列中获取 InnerEvent 事件或者 Runnable 任务。

（3）处理事件或任务。

如果 EventRunner 取出的事件为 InnerEvent，则触发 EventHandler 的回调方法并触发 EventHandler 的处理方法，在新线程上处理该事件。如果 EventRunner 取出的事件为 Runnable，则 EventRunner 直接在新线程上处理 Runnable 任务。

在进行线程间通信时，EventHandler 只能和 EventRunner 所创建的线程进行绑定，EventRunner 创建时需要判断是否创建成功，只有确保获取的 EventRunner 实例非空时，才可以使用 EventHandler 绑定 EventRunner。一个 EventHandler 只能同时与一个 EventRunner 绑定，一个 EventRunner 可以同时绑定多个 EventHandler。

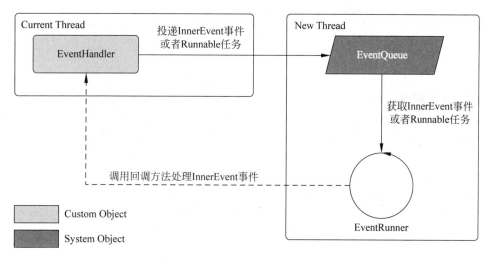

图 3-14 EventHandler 的运作机制

3.6.2 线程间接口关系

本节介绍 EventHandler、EventRunner 和 InnerEvent 3 种应用开发接口。

1. EventHandler

对于 EventHandler 开发场景,主要功能是将 InnerEvent 事件或者 Runnable 任务投递到其他的线程进行处理,使用场景如下。

InnerEvent 事件:需要将 InnerEvent 事件投递到新的线程,按照优先级和延时进行处理。投递时,EventHandler 的优先级可在 IMMEDIATE、HIGH、LOW 和 IDLE 中选择,并设置合适的 delayTime。Runnable 任务:需要将它投递到新的线程,并按照优先级和延时进行处理。投递时,EventHandler 的优先级可在 IMMEDIATE、HIGH、LOW 和 IDLE 中选择,并设置合适的 delayTime。还需要在新创建的线程中投递事件到原线程进行处理。

EventRunner 将根据优先级的高低从事件队列中获取事件或者 Runnable 任务进行处理。EventHandler 的属性 Priority(优先级)如表 3-20 所示。

表 3-20 EventHandler 的属性及功能描述

属　　性	功　能　描　述
Priority.IMMEDIATE	表示事件被立即投递
Priority.HIGH	表示事件先于 LOW 优先级投递
Priority.LOW	表示事件优先于 IDLE 优先级投递,事件的默认优先级是 LOW
Priority.IDLE	表示在没有其他事件的情况下,才投递该事件

EventHandler 的主要接口及功能描述如表 3-21 所示。

表 3-21 EventHandler 的主要接口及功能描述

主要接口	功能描述
EventHandler(EventRunner runner)	利用已有的 EventRunner 创建 EventHandler
current()	在 processEvent 回调中,获取当前的 EventHandler
processEvent(InnerEvent event)	回调处理事件,由开发者实现
sendEvent(InnerEvent event)	发送一个事件到事件队列,延时为 0ms,优先级为 LOW
sendEvent(InnerEvent event,long delayTime)	发送一个延时事件到事件队列,优先级为 LOW
sendEvent(InnerEvent event, long delayTime, EventHandler.Priority priority)	发送一个指定优先级的延时事件到事件队列
sendEvent(InnerEvent event,EventHandler.Priority priority)	发送一个指定优先级的事件到事件队列,延时为 0ms
sendSyncEvent(InnerEvent event)	发送一个同步事件到事件队列,延时为 0ms,优先级为 LOW
sendSyncEvent(InnerEvent event,EventHandler.Priority priority)	发送一个指定优先级的同步事件到事件队列,延时为 0ms,优先级不能是 IDLE
postSyncTask(Runnable task)	发送一个 Runnable 同步任务到事件队列,延时为 0ms,优先级为 LOW
postSyncTask(Runnable task,EventHandler.Priority priority)	发送一个指定优先级的 Runnable 同步任务到事件队列,延时为 0ms
postTask(Runnable task)	发送一个 Runnable 任务到事件队列,延时为 0ms,优先级为 LOW
postTask(Runnable task,long delayTime)	发送一个 Runnable 延时任务到事件队列,优先级为 LOW
postTask(Runnable task, long delayTime, EventHandler.Priority priority)	发送一个指定优先级的 Runnable 延时任务到事件队列
postTask(Runnable task,EventHandler.Priority priority)	发送一个指定优先级的 Runnable 任务到事件队列,延时为 0ms
sendTimingEvent(InnerEvent event,long taskTime)	发送一个定时事件到队列,在 taskTime 时间执行,如果 taskTime 小于当前时间,立即执行,优先级为 LOW
sendTimingEvent(InnerEvent event,long taskTime, EventHandler.Priority priority)	发送一个带优先级的事件到队列,在 taskTime 时间执行,如果 taskTime 小于当前时间,立即执行
postTimingTask(Runnable task,long taskTime)	发送一个 Runnable 任务到队列,在 taskTime 时间执行,如果 taskTime 小于当前时间,立即执行,优先级为 LOW
postTimingTask(Runnable task,long taskTime, EventHandler.Priority priority)	发送一个带优先级的 Runnable 任务到队列,在 taskTime 时间执行,如果 taskTime 小于当前时间,立即执行
removeEvent(int eventId)	删除指定 ID 的事件
removeEvent(int eventId,long param)	删除指定 ID 和 param 的事件

续表

主要接口	功能描述
removeEvent(int eventId, long param, Object object)	删除指定 ID、param 和 object 的事件
removeAllEvent()	删除 EventHandler 的所有事件
getEventName(InnerEvent event)	获取事件的名称
getEventRunner()	获取 EventHandler 绑定的 EventRunner
isIdle()	判断队列是否为空
hasInnerEvent(Runnable runnable)	根据指定的 runnable 参数,检查是否有还未被处理的任务。可以根据不同的入参进行检查,详见 EventHandler

2. EventRunner

EventRunner 工作模式可以分为托管模式和手动模式。两种模式是在调用 EventRunner 的 create()方法时,通过选择不同的参数实现。默认为托管模式。托管模式:不需要调用 run()和 stop()方法启动、停止 EventRunner。当 EventRunner 实例化时,系统调用 run()启动 EventRunner;当 EventRunner 不被引用时,系统调用 stop()停止 EventRunner。手动模式:需要自行调用 EventRunner 的 run()方法和 stop()方法确保线程的启动、停止。EventRunner 的主要接口及功能描述如表 3-22 所示。

表 3-22 EventRunner 的主要接口及功能描述

主要接口	功能描述
create()	创建一个拥有新线程的 EventRunner
create(boolean inNewThread)	创建一个拥有新线程的 EventRunner, inNewThread 为 true 时,EventRunner 为托管模式,系统将自动管理 EventRunner; inNewThread 为 false 时,EventRunner 为手动模式
create(String newThreadName)	创建一个拥有新线程的 EventRunner,新线程的名称是 newThreadName
current()	获取当前线程的 EventRunner
run()	EventRunner 为手动模式时,调用该方法启动新线程
stop()	EventRunner 为手动模式时,调用该方法停止新线程

3. InnerEvent

InnerEvent 的属性及功能描述如表 3-23 所示。

表 3-23 InnerEvent 的属性及功能描述

属性	功能描述
eventId	事件的 ID,由开发者定义用来辨别事件
object	事件携带的 Object 信息
param	事件携带的 long 型数据

InnerEvent 的主要接口及功能描述如表 3-24 所示。

表 3-24　InnerEvent 的主要接口及功能描述

主 要 接 口	功 能 描 述
drop()	释放一个事件实例
get()	获得一个事件实例
get(int eventId)	获得一个指定的 eventId 的事件实例
get(int eventId,long param)	获得一个指定的 eventId 和 param 的事件实例
get(int eventId,long param,Object object)	获得一个指定的 eventId、param 和 object 的事件实例
get(int eventId,Object object)	获得一个指定的 eventId 和 object 的事件实例
PacMap getPacMap()	获取 PacMap,如果没有,则新建一个
Runnable getTask()	获取 Runnable 任务
PacMap peekPacMap()	获取 PacMap
void setPacMap(PacMap pacMap)	设置 PacMap

3.6.3　线程间通信开发步骤

本节介绍 EventHandler 投递 InnerEvent 事件、EventHandler 投递 Runnable 任务和在新创建线程中投递事件到原线程。

1. EventHandler 投递 InnerEvent 事件

EventHandler 投递 InnerEvent 事件,并按照优先级和延时进行处理,开发步骤如下。

（1）创建 EventHandler 的子类,在子类中重写实现方法 processEvent()处理事件。

```
private static final int EVENT_MESSAGE_NORMAL = 1;
private static final int EVENT_MESSAGE_DELAY = 2;
private class MyEventHandler extends EventHandler {
    private MyEventHandler(EventRunner runner) {
        super(runner);
    }
    //重写实现 processEvent()方法
    @Override
    public void processEvent(InnerEvent event) {
        super.processEvent(event);
        if (event == null) {
            return;
        }
        int eventId = event.eventId;
        switch (eventId) {
            case EVENT_MESSAGE_NORMAL:
                //待执行的操作,由开发者定义
                break;
            case EVENT_MESSAGE_DELAY:
                //待执行的操作,由开发者定义
```

```
            break;
        default:
            break;
    }
}
```

(2) 以手动模式为例,创建 EventRunner。

```
EventRunner runner = EventRunner.create(false);
//create()的参数是 true 时,则为托管模式
```

(3) 创建 EventHandler 子类的实例。

```
MyEventHandler myHandler = new MyEventHandler(runner);
```

(4) 获取 InnerEvent 事件。

```
//获取事件实例,其属性 eventId、param 和 object 由开发者确定,代码中只是示例
long param = 0L;
Object object = null;
InnerEvent normalInnerEvent = InnerEvent.get(EVENT_MESSAGE_NORMAL, param, object);
InnerEvent delayInnerEvent = InnerEvent.get(EVENT_MESSAGE_DELAY, param, object);
```

(5) 投递事件:投递的优先级以 IMMEDIATE 为例,延时选择 0ms 和 2ms。

```
//优先级 IMMEDIATE,投递之后立即处理,延时为 0ms,该语句等价于同步投递 sendSyncEvent(event1,
EventHandler.Priority.IMMEDIATE);
myHandler.sendEvent(normalInnerEvent, 0, EventHandler.Priority.IMMEDIATE);
myHandler.sendEvent(delayInnerEvent, 2, EventHandler.Priority.IMMEDIATE);
//延时 2ms 后立即处理
```

(6) 启动和停止 EventRunner,如果为托管模式,则不需要此步骤。

```
runner.run();
//待执行操作
runner.stop();                    //开发者根据业务需要在适当时机停止 EventRunner
```

2. EventHandler 投递 Runnable 任务

EventHandler 投递 Runnable 任务,并按照优先级和延时进行处理,开发步骤如下。

(1) 创建 EventHandler 的子类,先创建 EventRunner,然后创建 EventHandler 子类的实例,步骤与 EventHandler 投递 InnerEvent 事件的步骤(1)~(3)相同。

(2) 创建 Runnable 任务。

```
Runnable normalTask = new Runnable() {
    @Override
    public void run() {
        //待执行的操作,由开发者定义
    }
```

```java
};
Runnable delayTask = new Runnable() {
    @Override
    public void run() {
        //待执行的操作,由开发者定义
    }
};
```

（3）投递 Runnable 任务,投递的优先级以 IMMEDIATE 为例,延时选择 0ms 和 2ms。

```java
//优先级为 immediate,延时 0ms,该语句等价于同步投递 myHandler.postSyncTask(task1,
EventHandler.Priority.IMMEDIATE);
myHandler.postTask(normalTask, 0, EventHandler.Priority.IMMEDIATE);
myHandler.postTask(delayTask, 2, EventHandler.Priority.IMMEDIATE);    //延时 2ms 后立即执行
```

（4）启动和停止 EventRunner,如果是托管模式,则不需要此步骤。

```java
runner.run();
//待执行操作
runner.stop();                        //停止 EventRunner
```

3. 在新创建线程中投递事件到原线程

EventHandler 从新创建线程投递事件到原线程并进行处理,开发步骤如下。

（1）创建 EventHandler 的子类,在子类中重写实现方法 processEvent()处理事件,相关代码如下。

```java
private static final int EVENT_MESSAGE_CROSS_THREAD = 1;
private class MyEventHandler extends EventHandler {
    private MyEventHandler(EventRunner runner) {
        super(runner);
    }
    //重写实现 processEvent()方法
    @Override
    public void processEvent(InnerEvent event) {
        super.processEvent(event);
        if (event == null) {
            return;
        }
        int eventId = event.eventId;
        switch (eventId) {
            case EVENT_MESSAGE_CROSS_THREAD:
                Object object = event.object;
                if (object instanceof EventRunner) {
                    //将原先线程的 EventRunner 实例投递给新创建的线程
                    EventRunner runner2 = (EventRunner) object;
                    //将原先线程的 EventRunner 实例与新创建线程的 EventHandler 绑定
                    EventHandler myHandler2 = new EventHandler(runner2) {
```

```
                @Override
                public void processEvent(InnerEvent event) {
                    //需要在原先线程执行的操作
                }
            };
            int eventId2 = 1;
            long param2 = 0L;
            Object object2 = null;
            InnerEvent event2 = InnerEvent.get(eventId2, param2, object2);
            myHandler2.sendEvent(event2);  //投递事件到原先的线程
        }
        break;
    default:
        break;
    }
  }
}
```

(2) 以手动模式为例,创建 EventRunner。

```
EventRunner runner = EventRunner.create(false);    // create()的参数是 true 时,则为托管模式
```

(3) 创建 EventHandler 子类的实例。

```
MyEventHandler myHandler = new MyEventHandler(runner);
```

(4) 获取 InnerEvent 事件。

```
//获取事件实例,其属性 eventId、param 和 object 由开发者确定,代码中只是示例
long param = 0L;
InnerEvent event = InnerEvent.get(EVENT_MESSAGE_CROSS_THREAD, param, EventRunner.current());
```

(5) 投递事件,在新线程上直接处理。

```
//将与当前线程绑定的 EventRunner 投递到与 runner 创建的新线程中
myHandler.sendEvent(event);
```

(6) 启动和停止 EventRunner,如果是托管模式,则不需要此步骤。

```
runner.run();                //待执行操作
runner.stop();               //停止 EventRunner
```

针对线程间通信,演示将 InnerEvent 事件投递到新的线程、将 Runnable 任务投递到新的线程、在新创建的线程投递事件到原线程进行处理。示例工程参考地址如下：https://gitee.com/openharmony/app_samples/tree/master/thread/EventHandler。

3.7 剪贴板开发

用户通过系统剪贴板服务,可实现应用之间的简单数据传递。例如,在应用 A 中复制

的数据,可以在应用 B 中粘贴,反之亦可。

HarmonyOS 提供系统剪贴板服务的操作接口,支持用户程序从系统剪贴板中读取、写入和查询剪贴板数据,以及添加、移除系统剪贴板数据变化的回调。HarmonyOS 提供的剪贴板数据的对象定义包含内容对象和属性对象。

3.7.1 剪贴板开发接口关系

同一设备的应用程序 A、B 之间可以借助系统剪贴板服务完成简单数据的传递,即应用程序 A 向剪贴板服务写入数据后,应用程序 B 可以从中读取数据,如图 3-15 所示。

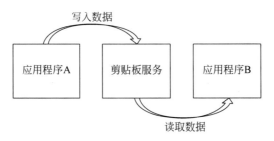

图 3-15 剪贴板服务示意图

在使用剪贴板服务时,需要注意以下几点:只有在前台获取到焦点的应用才有读取系统剪贴板的权限(系统默认输入法应用除外)。写入剪贴板服务中的数据不会随应用程序结束而销毁。对同一用户而言,写入剪贴板服务的数据会被下次写入的剪贴板数据覆盖。在同一设备内,剪贴板单次传递内容不应超过 500KB。

SystemPasteboard 提供系统剪贴板操作的相关接口包括复制、粘贴、配置回调等。PasteData 是剪贴板服务操作的数据对象,一个 PasteData 由若干个内容节点(PasteData.Record)和一个属性集合对象(PasteData.DataProperty)组成。Record 是存放剪贴板数据内容信息的最小单位,每个 Record 都有其特定的 MIME 类型,例如纯文本、HTML、URI 和 Intent。剪贴板数据的属性信息存在放 PasteData.DataProperty 中,包括标签、时间戳等。

1. SystemPasteboard

SystemPasteboard 的主要接口及功能描述如表 3-25 所示。

表 3-25 SystemPasteboard 的主要接口及功能描述

主要接口	功能描述
getSystemPasteboard(Context context)	获取系统剪贴板服务的对象实例
getPasteData()	读取当前系统剪贴板中的数据
hasPasteData()	判断当前系统剪贴板中是否有内容
setPasteData(PasteData data)	将剪贴板数据写入系统剪贴板
clear()	清空系统剪贴板数据

续表

主要接口	功能描述
addPasteDataChangedListener (IPasteDataChangedListener listener)	用户程序添加系统剪贴板数据变化的回调,当系统剪贴板数据发生变化时,会触发用户程序的回调实现
removePasteDataChangedListener (IPasteDataChangedListener listener)	用户程序移除系统剪贴板数据变化的回调

2. PasteData

PasteData 是剪贴板服务操作的数据对象,其中内容节点定义为 PasteData.Record,属性集合定义为 PasteData.DataProperty,如表 3-26 所示。

表 3-26　PasteData 的主要接口及功能描述

主要接口	功能描述
PasteData()	构造器,创建一个空内容数据对象
creatPlainTextData(CharSequence text)	构建一个包含纯文本内容节点的数据对象
creatHtmlData(String htmlText)	构建一个包含 HTML 内容节点的数据对象
creatUriData(Uri uri)	构建一个包含 URI 内容节点的数据对象
creatIntentData(Intent intent)	构建一个包含 Intent 内容节点的数据对象
getPrimaryMimeType()	获取数据对象中首个内容节点的 MIME 类型,如果没有查询到内容,将返回一个空字符串
getPrimaryText()	获取数据对象中首个内容节点的纯文本内容,如果没有查询到内容,将返回一个空对象
addTextRecord(CharSequence text)	向数据对象中添加一个纯文本内容节点,该方法会自动更新数据属性中的 MIME 类型集合,最多只能添加 128 个内容节点
addRecord(Record record)	向数据对象中添加一个内容节点,该方法会自动更新数据属性中的 MIME 类型集合,最多只能添加 128 个内容节点
getRecordCount()	获取数据对象中内容节点的数量
getRecordAt(int index)	获取数据对象在指定下标处的内容节点,如果操作失败会返回空对象
removeRecordAt(int index)	移除数据对象在指定下标处的内容节点,如果操作成功会返回 true,操作失败会返回 false
getMimeTypes()	获取数据对象中所有内容节点的 MIME 类型列表,当内容节点为空时,返回列表为空对象
getProperty()	获取该数据对象的属性集合成员

PasteData 中定义的常量名及功能描述如表 3-27 所示。

表 3-27　PasteData 中定义的常量名及功能描述

常　量　名	功　能　描　述
MIMETYPE_TEXT_PLAIN= "text/plain"	纯文本的 MIME 类型定义
MIMETYPE_TEXT_HTML= "text/html"	HTML 的 MIME 类型定义
MIMETYPE_TEXT_URI= "text/uri"	URI 的 MIME 类型定义
MIMETYPE_TEXT_INTENT= "text/ohos.intent"	Intent 的 MIME 类型定义
MAX_RECORD_NUM=128	单个 PasteData 中所能包含的 Record 的数量上限

3. PasteData.Record

一个 PasteData 中包含若干个特定 MIME 类型的 PasteData.Record，每个 Record 是存放剪贴板数据内容信息的最小单位，如表 3-28 所示。

表 3-28　PasteData.Record 的主要接口及功能描述

主　要　接　口	功　能　描　述
createPlainTextRecord(CharSequence text)	构造一个 MIME 类型为纯文本的内容节点
createHtmlTextRecord(String htmlText)	构造一个 MIME 类型为 HTML 的内容节点
createUriRecord(Uri uri)	构造一个 MIME 类型为 URI 的内容节点
createIntentRecord(Intent intent)	构造一个 MIME 类型为 Intent 的内容节点
getPlainText()	获取该内容节点中的文本内容，如果没有内容返回空对象
getHtmlText()	获取该内容节点中的 HTML 内容，没有内容将返回空对象
getUri()	获取该内容节点中的 URI 内容，如果没有内容返回空对象
getIntent()	获取该内容节点中的 Intent 内容，如果没有内容返回空对象
getMimeType()	获取该内容节点的 MIME 类型
convertToText(Context context)	将该内容节点的内容转为文本形式

4. PasteData.DataProperty

每个 PasteData 中都有一个 PasteData.DataProperty 成员，存放该数据对象的属性集合，例如自定义标签、MIME 类型集合列表等，主要接口及功能描述如表 3-29 所示。

表 3-29　PasteData.DataProperty 的主要接口及功能描述

主　要　接　口	功　能　描　述
getMimeTypes()	获取所属数据对象的 MIME 类型集合列表，当内容节点为空时，返回列表为空对象
hasMimeType(String mimeType)	判断所属数据对象中是否包含特定 MIME 类型的内容
getTimestamp()	获取所属数据对象被写入系统剪贴板时的时间戳，如果该数据对象尚未被写入，则返回 0
setTag(CharSequence tag)	设置自定义标签
getTag()	获取自定义标签

续表

主要接口	功能描述
setAdditions(PacMap extraProps)	设置一些附加键值对信息
getAdditions()	获取附加键值对信息

5. IPasteDataChangedListener

IPasteDataChangedListener 是定义剪贴板数据变化回调的接口类,开发者需要实现此接口编码触发回调时的处理逻辑,主要接口及功能描述如表 3-30 所示。

表 3-30　IPasteDataChangedListener 的主要接口及功能描述

主要接口	功能描述
onChanged()	系统剪贴板数据发生变化时的回调接口

3.7.2　剪贴板开发步骤

剪贴板的主要开发步骤如下。

(1) 应用 A 获取系统剪贴板服务,相关代码如下。

```java
SystemPasteboard pasteboard = SystemPasteboard.getSystemPasteboard(appContext);
```

(2) 应用 A 向系统剪贴板中写入一条纯文本数据,相关代码如下。

```java
if (pasteboard != null) {
    pasteboard.setPasteData(PasteData.creatPlainTextData("Hello, world!"));
}
```

(3) 应用 B 从系统剪贴板读取数据,将数据对象中的首个文本类型(纯文本/HTML)内容信息在控件中显示,忽略其他类型内容,相关代码如下。

```java
PasteData pasteData = pasteboard.getPasteData();
if (pasteData == null) {
    return;
}
DataProperty dataProperty = pasteData.getProperty();
boolean hasHtml = dataProperty.hasMimeType(PasteData.MIMETYPE_TEXT_HTML);
boolean hasText = dataProperty.hasMimeType(PasteData.MIMETYPE_TEXT_PLAIN);
if (hasHtml || hasText) {
    Text text = (Text) findComponentById(ResourceTable.Id_text);
    for (int i = 0; i < pasteData.getRecordCount(); i++) {
        PasteData.Record record = pasteData.getRecordAt(i);
        String mimeType = record.getMimeType();
        if (mimeType.equals(PasteData.MIMETYPE_TEXT_HTML)) {
            text.setText(record.getHtmlText());
            break;
```

```
            } else if (mimeType.equals(PasteData.MIMETYPE_TEXT_PLAIN)) {
                text.setText(record.getPlainText().toString());
                break;
            } else {
                //跳过其他 Mime 类型的记录
            }
        }
    }
```

(4)应用 C 注册添加系统剪贴板数据变化回调,当系统剪贴板数据发生变化时触发处理逻辑,相关代码如下。

```
IPasteDataChangedListener listener = new IPasteDataChangedListener() {
    @Override
    public void onChanged() {
        PasteData pasteData = pasteboard.getPasteData();
        if (pasteData == null) {
            return;
        }
        //处理系统剪贴板上数据更改的操作
    }
};
pasteboard.addPasteDataChangedListener(listener);
```

针对剪贴板开发指导,演示应用之间的数据剪贴。示例工程参考地址为:https://gitee.com/openharmony/app_samples/tree/master/ability/Pasteboard。

第 4 章 Java UI 开发

本章主要对组件与布局开发、常用布局开发、自定义组件与布局、动画开发及可视即可说开发进行介绍。

4.1 Java UI 框架概述

应用的 Ability 在屏幕上将显示一个用户界面,该界面显示所有可被用户查看和交互的内容。

应用中所有的用户界面元素都是由 Component 和 ComponentContainer 对象构成的。Component 是绘制在屏幕上的一个对象,用户能与之交互。ComponentContainer 是一个用于容纳其他 Component 和 ComponentContainer 对象的容器。

Java UI 框架提供了一部分 Component 和 ComponentContainer 的具体子类,即创建用户界面的各类组件,包括一些常用的组件(如文本、按钮、图片、列表等)和常用的布局(如 DirectionalLayout 和 DependentLayout)。用户可通过组件进行交互操作,并获得响应,所有的 UI 操作都应该在主线程进行设置。

用户界面元素统称为组件,组件根据一定的层级结构进行组合形成布局。组件在未被添加到布局时,既无法显示也无法交互,因此一个用户界面至少包含一个布局。在 UI 框架中,具体的布局类通常以 XXLayout 命名,完整的用户界面是一个布局,用户界面中的一部分也可以是一个布局,布局中容纳 Component 与 ComponentContainer 对象。

Component:提供内容显示,Component 是界面中所有组件的基类,开发者可以给 Component 设置事件处理回调创建一个可交互的组件。Java UI 框架提供了一些常用的界面元素,也可称之为组件,组件一般直接继承 Component 或它的子类,例如 Text、Image 等。

ComponentContainer:作为容器容纳 Component 或 ComponentContainer 对象,并对它们进行布局。Java UI 框架提供了一些标准布局功能的容器,它们继承自 ComponentContainer,一般以 Layout 结尾,例如 DirectionalLayout、DependentLayout 等,如图 4-1 所示。

每种布局都根据自身特点提供 LayoutConfig,供子 Component 设定布局属性和参数,

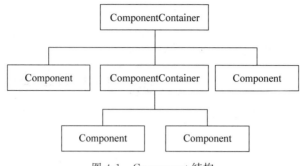

图 4-1 Component 结构

通过指定布局属性可以对子 Component 在布局中的显示效果进行约束。例如，width、height 是最基本的布局属性，它们指定了组件的大小，LayoutConfig 布局如图 4-2 所示。

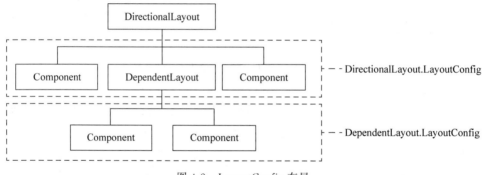

图 4-2 LayoutConfig 布局

布局把 Component 和 ComponentContainer 以树状的层级结构进行组织，这样的一个布局称为组件树。组件树的特点是仅有一个根组件，其他组件有且仅有一个父节点，组件之间的关系受到父节点的规则约束。

4.2 组件与布局开发

HarmonyOS 提供了 Ability 和 AbilitySlice 两个基础类，一个有界面的 Ability 可以由一个或多个 AbilitySlice 构成，AbilitySlice 主要用于承载单个页面的具体逻辑实现和界面 UI，是应用显示、运行和跳转的最小单元。AbilitySlice 通过 setUIContent 为界面设置布局，主要接口及功能描述如表 4-1 所示。

表 4-1 AbilitySlice 的 UI 主要接口及功能描述

主 要 接 口	功 能 描 述
setUIContent(ComponentContainer root)	设置界面入口，root 为界面组件树根节点

组件需要进行组合,并添加到界面的布局中。Java UI 框架中,提供了两种编写布局的方式。

(1) 在代码中创建布局,用代码创建 Component 和 ComponentContainer 对象,为这些对象设置合适的参数和属性值,并将 Component 添加到 ComponentContainer 中,从而创建出完整界面。

(2) 在 XML 中声明 UI 布局,按层级结构描述 Component 和 ComponentContainer 的关系,给组件节点设定合适的参数和属性值,代码中可直接加载生成此布局。这两种方式创建出的布局没有本质差别,在 XML 中声明布局,加载后同样可以对代码中的布局进行修改。

根据组件的功能,可以分为布局类、显示类、交互类,如表 4-2 所示。

表 4-2 组件分类及功能描述

组件类别	组件名称	功能描述
布局类	PositionLayout、DirectionalLayout、StackLayout、DependentLayout、TableLayout、AdaptiveBoxLayout	提供不同布局规范的组件容器,例如以单一方向排列的 DirectionalLayout、以相对位置排列的 DependentLayout、以确切位置排列的 PositionLayout 等
显示类	Text、Image、Clock、TickTimer、ProgressBar	提供单纯的内容显示,例如用于文本显示的 Text、用于图像显示的 Image 等
交互类	TextField、Button、Checkbox、RadioButton/RadioContainer、Switch、ToggleButton、Slider、Rating、ScrollView、TabList、ListContainer、PageSlider、PageFlipper、PageSliderIndicator、Picker、TimePicker、DatePicker、SurfaceProvider、ComponentProvider	提供具体场景下与用户交互响应的功能,例如 Button 提供了单击响应功能,Slider 提供了进度选择功能,等等

1. 代码创建布局

开发样例如图 4-3 所示,需要添加一个 Text 组件和一个 Button 组件。由于两个组件从上到下依次居中排列,因此可以选择使用竖向的 DirectionalLayout 布局放置组件。

代码创建布局需要在 MainAbilitySlice 中分别创建组件和布局,并将它们进行关联,文件目录如图 4-4 所示。

My name is Text.

My name is Button.

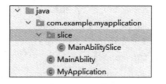

图 4-3 开发样例图　　　　　　　　图 4-4 文件目录

1) 创建组件

创建组件步骤如下。

(1) 声明组件。

```
Button button = new Button(getContext());
```

(2) 设置组件大小。

```
button.setWidth(ComponentContainer.LayoutConfig.MATCH_CONTENT);
button.setHeight(ComponentContainer.LayoutConfig.MATCH_CONTENT);
```

(3) 设置组件属性。

```
button.setText("My name is Button.");
button.setTextSize(50);
```

2) 创建布局并使用

创建布局步骤如下。

(1) 声明布局。

```
DirectionalLayout directionalLayout = new DirectionalLayout(getContext());
```

(2) 设置布局大小。

```
directionalLayout.setWidth(ComponentContainer.LayoutConfig.MATCH_PARENT);
directionalLayout.setHeight(ComponentContainer.LayoutConfig.MATCH_PARENT);
```

(3) 设置布局属性。

```
directionalLayout.setOrientation(Component.VERTICAL);
```

(4) 将组件添加到布局中(视布局需要对组件设置属性进行约束)。

```
directionalLayout.addComponent(button);
```

(5) 将布局添加到组件树中。

```
setUIContent(directionalLayout);
```

3) 示例代码

```
//请根据实际工程/包名情况引入
package com.example.myapplication.slice;
import ohos.aafwk.ability.AbilitySlice;
import ohos.aafwk.content.Intent;
import ohos.agp.colors.RgbColor;
import ohos.agp.components.*;
import ohos.agp.components.element.ShapeElement;
import ohos.agp.utils.LayoutAlignment;
public class MainAbilitySlice extends AbilitySlice {
```

```java
@Override
public void onStart(Intent intent) {
    super.onStart(intent);
    //声明布局
    DirectionalLayout directionalLayout = new DirectionalLayout(getContext());
    //设置布局大小
    directionalLayout.setWidth(ComponentContainer.LayoutConfig.MATCH_PARENT);
    directionalLayout.setHeight(ComponentContainer.LayoutConfig.MATCH_PARENT);
    //设置布局属性
    directionalLayout.setOrientation(Component.VERTICAL);
    directionalLayout.setPadding(32, 32, 32, 32);
    Text text = new Text(getContext());
    text.setText("My name is Text.");
    text.setTextSize(50);
    text.setId(100);
    //为组件添加对应的布局属性
    DirectionalLayout.LayoutConfig layoutConfig = new DirectionalLayout.LayoutConfig
(ComponentContainer.LayoutConfig.MATCH_CONTENT, ComponentContainer.LayoutConfig.MATCH_
CONTENT);
    layoutConfig.alignment = LayoutAlignment.HORIZONTAL_CENTER;
    text.setLayoutConfig(layoutConfig);
    //将 Text 添加到布局中
    directionalLayout.addComponent(text);
    //添加一个 Button
    Button button = new Button(getContext());
    layoutConfig.setMargins(0, 50, 0, 0);
    button.setLayoutConfig(layoutConfig);
    button.setText("My name is Button.");
    button.setTextSize(50);
    ShapeElement background = new ShapeElement();
    background.setRgbColor(new RgbColor(0, 125, 255));
    background.setCornerRadius(25);
    button.setBackground(background);
    button.setPadding(10, 10, 10, 10);
    button.setClickedListener(new Component.ClickedListener() {
        @Override
        //在组件中增加对单击事件的检测
        public void onClick(Component component) {
            //此处添加按钮被单击需要执行的操作
        }
    });
    directionalLayout.addComponent(button);
    //将布局作为根布局添加到视图树中
    super.setUIContent(directionalLayout);
}
}
```

代码示例中为组件设置了一个按键回调,在按键被按下后,应用会执行自定义的操作。

在代码示例中,可以看到设置组件大小的方法有两种:①通过 setWidth/setHeight 直接设置宽高;②通过 setLayoutConfig 方法设置布局属性来设定宽高。

这两种方法的区别是,后者还可以增加更多的布局属性设置,例如使用 alignment 设置水平居中的约束。另外,这两种方法设置的宽高以最后的设置作为最终结果。它们的取值一致,以像素为单位的数值。MATCH_PARENT:表示组件大小将扩展为父组件允许的最大值,它将占据父组件方向上的剩余大小。MATCH_CONTENT:表示组件大小与它内容占据的大小范围相适应。

2. XML 创建布局

XML 声明布局的方式更加简便直观。每个 Component 和 ComponentContainer 对象大部分属性都支持在 XML 中进行设置,它们都有各自的 XML 属性列表。某些属性仅适用于特定的组件。例如,只有 Text 支持 text_color 属性,但不支持该属性的组件,如果添加了该属性,则会被忽略。具有继承关系的组件子类将继承父类的属性列表,Component 作为组件的基类,拥有各个组件常用的属性,例如 ID、布局参数等。

ID 设置为 ohos:id="$+id:text"。

在 XML 中使用此格式声明一个对开发者友好的 ID,它会在编译过程中转换成一个常量。尤其在 DependentLayout 布局中,组件之间需要描述相对位置关系,描述时要通过 ID 指定对应组件。布局中的组件通常要设置独立的 ID,以便在程序中查找该组件。如果布局中有不同组件设置了相同的 ID,在通过 ID 查找组件时会返回查找到的第一个组件,因此尽量保证在所要查找的布局中为组件设置独立的 ID 值,避免出现与预期不相符的问题。

布局参数设置如下。

```
ohos:width = "20vp"
ohos:height = "10vp"
```

与代码中设置组件的宽度和高度类似,XML 中具体数值:10(以像素为单位)、10vp(以屏幕相对像素为单位)。match_parent:表示组件大小将扩展为父组件允许的最大值,它将占据父组件方向上的剩余大小。match_content:表示组件大小与内容占据的大小范围相适应。

1) 创建 XML 布局文件

创建步骤如下。

(1) 在 DevEco Studio 的 Project 窗口中,打开 entry→src→main→resources→base,右击 layout 文件夹,选择 New→Layout Resource File,命名为 first_layout,如图 4-5 所示。

(2) 打开新创建的 first_layout.xml 布局文件,修改其中的内容,对布局、组件的属性和层级进行描述,相关代码如下。

```
<?xml version = "1.0" encoding = "utf-8"?>
< DirectionalLayout
```

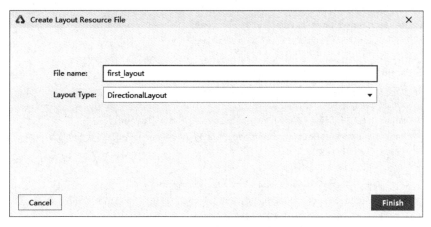

图 4-5　创建 XML 布局

```
xmlns:ohos = "http://schemas.huawei.com/res/ohos"
ohos:width = "match_parent"
ohos:height = "match_parent"
ohos:orientation = "vertical"
ohos:padding = "32">
< Text
    ohos:id = " $ + id:text"
    ohos:width = "match_content"
    ohos:height = "match_content"
    ohos:layout_alignment = "horizontal_center"
    ohos:text = "My name is Text."
    ohos:text_size = "25fp"/>
< Button
    ohos:id = " $ + id:button"
    ohos:margin = "50"
    ohos:width = "match_content"
    ohos:height = "match_content"
    ohos:layout_alignment = "horizontal_center"
    ohos:text = "My name is Button."
    ohos:text_size = "50"/>
</DirectionalLayout >
```

2) 加载 XML 布局

在代码中需要加载 XML 布局，并添加为根布局或作为其他布局的子 Component。

```
//根据实际工程/包名情况引入
package com.example.myapplication.slice;
import com.example.myapplication.ResourceTable;
import ohos.aafwk.ability.AbilitySlice;
import ohos.aafwk.content.Intent;
import ohos.agp.colors.RgbColor;
```

```
import ohos.agp.components.*;
import ohos.agp.components.element.ShapeElement;
public class ExampleAbilitySlice extends AbilitySlice {
    @Override
    public void onStart(Intent intent) {
        super.onStart(intent);
        //加载 XML 布局作为根布局
        super.setUIContent(ResourceTable.Layout_first_layout);
        Button button = (Button) findComponentById(ResourceTable.Id_button);
        if (button != null) {
            //设置组件的属性
            ShapeElement background = new ShapeElement();
            background.setRgbColor(new RgbColor(0, 125, 255));
            background.setCornerRadius(25);
            button.setBackground(background);
            button.setClickedListener(new Component.ClickedListener() {
                @Override
                //在组件中增加对单击事件的检测
                public void onClick(Component component) {
                    //此处添加按钮被单击需要执行的操作
                }
            });
        }
    }
}
```

4.3 常用组件开发

常用组件开发包括组件通用 XML 属性、Text、Button、TextField、Image、TabList、Tab、Picker、DatePicker、TimePicker、Switch、RadioButton、RadioContainer、Checkbox、ProgressBar、RoundProgressBar、ToastDialog、PopupDialog、CommonDialog、ScrollView、ListContainer、PageSlider、PageSliderIndicator 和 WebView。

4.3.1 组件通用 XML 属性

Component 是所有组件的基类，Component 支持的 XML 属性，其他组件全部支持。Component 支持的 XML 属性请扫描二维码获取。

4.3.2 Text

Text 是显示字符串的组件，在界面上显示为一块文本区域。Text 作为一个基本组件，有很多扩展，常见的有按钮组件 Button 和文本编辑组件 TextField。Text 的共有 XML 属性继承自 Component，Text 的自有 XML 属性请扫描二维码获取。

1. 创建 Text

在 layout 目录下的 XML 文件中创建 Text 组件,相关代码如下。

```xml
<Text
    ohos:id = "$+id:text"
    ohos:width = "match_content"
    ohos:height = "match_content"
    ohos:text = "Text"/>
```

2. 设置 Text

设置 Text 步骤如下。

(1) 在 XML 中设置 Text 的背景。

layout 目录下 XML 文件的相关代码如下。

```xml
<Text
    …
    ohos:background_element = "$graphic:background_text"/>
```

常用的背景如常见的文本背景、按钮背景,可以采用 XML 格式放置在 graphic 目录下。

在 Project 窗口中,打开 entry→src→main→resources→base,右击 graphic 文件夹,选择 New→File,命名为 background_text.xml,在 background_text.xml 中定义文本的背景,相关代码如下。

```xml
<?xml version = "1.0" encoding = "utf-8"?>
<shape xmlns:ohos = "http://schemas.huawei.com/res/ohos"
    ohos:shape = "rectangle">
    <corners
        ohos:radius = "20"/>
    <solid
        ohos:color = "#878787"/>
</shape>
```

(2) 设置字体大小和颜色,相关代码如下。

```xml
<Text
    ohos:id = "$+id:text"
    ohos:width = "match_content"
    ohos:height = "match_content"
    ohos:text = "Text"
    ohos:text_size = "28fp"
    ohos:text_color = "#0000FF"
    ohos:left_margin = "15vp"
    ohos:bottom_margin = "15vp"
    ohos:right_padding = "15vp"
    ohos:left_padding = "15vp"
    ohos:background_element = "$graphic:background_text"/>
```

(3) 设置字体风格和字重,相关代码如下。

```xml
< Text
    ohos:id = " $ + id:text"
    ohos:width = "match_content"
    ohos:height = "match_content"
    ohos:text = "Text"
    ohos:text_size = "28fp"
    ohos:text_color = " #0000FF"
    ohos:italic = "true"
    ohos:text_weight = "700"
    ohos:text_font = "serif"
    ohos:left_margin = "15vp"
    ohos:bottom_margin = "15vp"
    ohos:right_padding = "15vp"
    ohos:left_padding = "15vp"
    ohos:background_element = " $graphic:background_text"/>
```

(4) 设置文本对齐方式,相关代码如下。

```xml
< Text
    ohos:id = " $ + id:text"
    ohos:width = "300vp"
    ohos:height = "100vp"
    ohos:text = "Text"
    ohos:text_size = "28fp"
    ohos:text_color = " #0000FF"
    ohos:italic = "true"
    ohos:text_weight = "700"
    ohos:text_font = "serif"
    ohos:left_margin = "15vp"
    ohos:bottom_margin = "15vp"
    ohos:right_padding = "15vp"
    ohos:left_padding = "15vp"
    ohos:text_alignment = "horizontal_center|bottom"
    ohos:background_element = " $graphic:background_text"/>
```

(5) 设置文本换行和最大显示行数,相关代码如下。

```xml
< Text
    ohos:id = " $ + id:text"
    ohos:width = "75vp"
    ohos:height = "match_content"
    ohos:text = "TextText"
    ohos:text_size = "28fp"
    ohos:text_color = " #0000FF"
    ohos:italic = "true"
    ohos:text_weight = "700"
```

```
ohos:text_font = "serif"
ohos:multiple_lines = "true"
ohos:max_text_lines = "2"
ohos:background_element = " $graphic:background_text"/>
```

3. 自动调节字体大小

Text 对象支持根据文本长度自动调整字体大小和换行。

(1) 设置自动换行、最大显示行数和自动调节字体大小，相关代码如下。

```
< Text
    ohos:id = " $ + id:text"
    ohos:width = "90vp"
    ohos:height = "match_content"
    ohos:min_height = "30vp"
    ohos:text = "T"
    ohos:text_color = " #0000FF"
    ohos:italic = "true"
    ohos:text_weight = "700"
    ohos:text_font = "serif"
    ohos:multiple_lines = "true"
    ohos:max_text_lines = "1"
    ohos:auto_font_size = "true"
    ohos:right_padding = "8vp"
    ohos:left_padding = "8vp"
    ohos:background_element = " $graphic:background_text"/>
```

(2) 通过 setAutoFontSizeRule 设置自动调整规则，三个入参分别是最小的字体大小、最大的字体大小、每次调整文本字体大小的步长，相关代码如下。

```
Text text = (Text) findComponentById(ResourceTable.Id_text);
//设置自动调整规则
text.setAutoFontSizeRule(30, 100, 1);
//设置单击一次增多一个"T"
text.setClickedListener(new Component.ClickedListener() {
    @Override
    public void onClick(Component component) {
        text.setText(text.getText() + "T");
    }
});
```

4. 跑马灯效果

当文本过长时，可以设置跑马灯效果，实现文本滚动显示。前提是文本换行关闭且最大显示行数为1，默认情况下即可满足前提要求，相关代码如下。

```
< Text
    ohos:id = " $ + id:text"
    ohos:width = "75vp"
```

```
        ohos:height = "match_content"
        ohos:text = "TextText"
        ohos:text_size = "28fp"
        ohos:text_color = "#0000FF"
        ohos:italic = "true"
        ohos:text_weight = "700"
        ohos:text_font = "serif"
        ohos:background_element = " $graphic:background_text"/>
//跑马灯效果
text.setTruncationMode(Text.TruncationMode.AUTO_SCROLLING);
//始终处于自动滚动状态
text.setAutoScrollingCount(Text.AUTO_SCROLLING_FOREVER);
//启动跑马灯效果
text.startAutoScrolling();
```

5. 场景示例

利用文本组件实现一个标题栏和详细内容的界面,如图 4-6 所示。

相关代码如下。

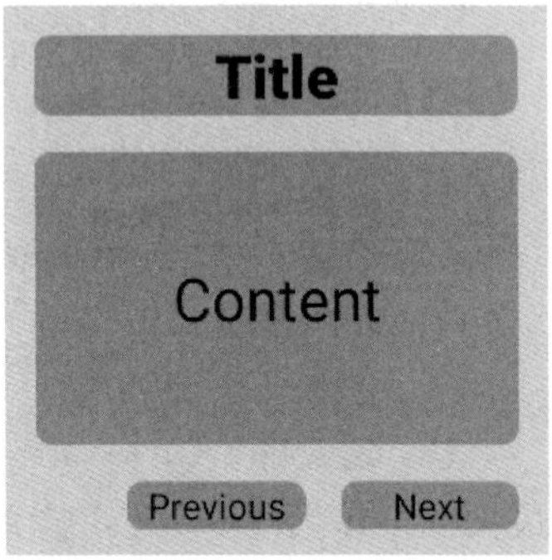

```
<?xml version = "1.0" encoding = "utf-8"?>
<DependentLayout
    xmlns:ohos = "http://schemas.huawei.com/res/ohos"
    ohos:width = "match_parent"
    ohos:height = "match_content"
    ohos:background_element = " $graphic:color_light_gray_element">
    <Text
        ohos:id = " $ + id:text1"
        ohos:width = "match_parent"
        ohos:height = "match_content"
        ohos:text_size = "25fp"
        ohos:top_margin = "15vp"
        ohos:left_margin = "15vp"
        ohos:right_margin = "15vp"
        ohos:background_element = " $graphic:background_text"
        ohos:text = "Title"
        ohos:text_weight = "1000"
        ohos:text_alignment = "horizontal_center"/>
    <Text
        ohos:id = " $ + id:text2"
        ohos:width = "match_parent"
        ohos:height = "120vp"
        ohos:text_size = "25fp"
        ohos:background_element = " $graphic:background_text"
        ohos:text = "Content"
        ohos:top_margin = "15vp"
        ohos:left_margin = "15vp"
```

图 4-6　界面效果

```
            ohos:right_margin = "15vp"
            ohos:bottom_margin = "15vp"
            ohos:text_alignment = "center"
            ohos:below = " $id:text1"
            ohos:text_font = "serif"/>
    < Button
            ohos:id = " $ + id:button1"
            ohos:width = "75vp"
            ohos:height = "match_content"
            ohos:text_size = "15fp"
            ohos:background_element = " $graphic:background_text"
            ohos:text = "Previous"
            ohos:right_margin = "15vp"
            ohos:bottom_margin = "15vp"
            ohos:left_padding = "5vp"
            ohos:right_padding = "5vp"
            ohos:below = " $id:text2"
            ohos:left_of = " $id:button2"
            ohos:text_font = "serif"/>
    < Button
            ohos:id = " $ + id:button2"
            ohos:width = "75vp"
            ohos:height = "match_content"
            ohos:text_size = "15fp"
            ohos:background_element = " $graphic:background_text"
            ohos:text = "Next"
            ohos:right_margin = "15vp"
            ohos:bottom_margin = "15vp"
            ohos:left_padding = "5vp"
            ohos:right_padding = "5vp"
            ohos:align_parent_end = "true"
            ohos:below = " $id:text2"
            ohos:text_font = "serif"/>
</DependentLayout >
color_light_gray_element.xml:
<?xml version = "1.0" encoding = "utf - 8"?>
< shape xmlns:ohos = "http://schemas.huawei.com/res/ohos"
        ohos:shape = "rectangle">
        < solid
            ohos:color = " # EDEDED"/>
</ shape >
background_text.xml:
<?xml version = "1.0" encoding = "utf - 8"?>
< shape xmlns:ohos = "http://schemas.huawei.com/res/ohos"
        ohos:shape = "rectangle">
        < corners
            ohos:radius = "20"/>
```

```
        <solid
            ohos:color = "#878787"/>
</shape>
```

4.3.3 Button

Button 是一种常见的组件，单击可以触发对应的操作，通常由文本或图标组成，也可以由图标和文本共同组成。Button 支持 XML 属性，无自有的 XML 属性，共有 XML 属性继承自 Text。

1. 创建 Button

创建 Button 步骤如下。

（1）在 layout 目录下的 XML 文件中创建 Button，并设置按钮的背景形状、颜色。常用的有文本背景、按钮背景，采用 XML 格式放置在 graphic 目录下，相关代码如下。

```
<Button
    ohos:id = "$+id:button"
    ohos:width = "match_content"
    ohos:height = "match_content"
    ohos:text_size = "27fp"
    ohos:text = "button"
    ohos:background_element = "$graphic:background_button"
    ohos:left_margin = "15vp"
    ohos:bottom_margin = "15vp"
    ohos:right_padding = "8vp"
    ohos:left_padding = "8vp"
    ohos:element_left = "$media:ic_btn_reload"
/>
```

（2）在 Project 窗口中，打开 entry→src→main→resources→base→media，添加所需图片至 media 目录下。

（3）在 Project 窗口中，打开 entry→src→main→resources→base，右击 graphic 文件夹，选择 New→File，命名为 background_button.xml，在该文件中定义按钮的背景形状、颜色，相关代码如下。

```
<?xml version = "1.0" encoding = "utf-8"?>
<shape xmlns:ohos = "http://schemas.huawei.com/res/ohos"
    ohos:shape = "rectangle">
    <corners
        ohos:radius = "10"/>
    <solid
        ohos:color = "#007CFD"/>
</shape>
```

2. 响应单击事件

按钮的重要作用是当用户单击按钮时，会执行相应的操作或者界面出现相应的变化。

实际上用户单击按钮时，Button 对象将收到一个单击事件。开发者可以自定义响应单击事件的方法。例如，通过创建一个 Component.ClickedListener 对象，调用 setClickedListener，将其分配给按钮，相关代码如下。

```
Button button = (Button) findComponentById(ResourceTable.Id_button);
//为按钮设置单击事件回调
button.setClickedListener(new Component.ClickedListener() {
    @Override
    public void onClick(Component component) {
        //此处添加单击按钮后的事件处理逻辑
    }
});
```

3. 不同类型的按钮

按钮按照形状可以分为普通按钮、椭圆按钮、胶囊按钮、圆形按钮等。

1）普通按钮

普通按钮和其他按钮的区别在于不需要设置任何形状，只设置文本和背景颜色即可，相关代码如下。

```xml
<Button
    ohos:width = "150vp"
    ohos:height = "50vp"
    ohos:text_size = "27fp"
    ohos:text = "button"
    ohos:background_element = "$graphic:color_blue_element"
    ohos:left_margin = "15vp"
    ohos:bottom_margin = "15vp"
    ohos:right_padding = "8vp"
    ohos:left_padding = "8vp"
/>
```

graphic 目录下的 color_blue_element.xml 文件示例如下。

```xml
<?xml version = "1.0" encoding = "utf-8"?>
<shape xmlns:ohos = "http://schemas.huawei.com/res/ohos"
    ohos:shape = "rectangle">
    <solid
        ohos:color = "#007CFD"/>
</shape>
```

2）椭圆按钮

椭圆按钮通过设置 background_element 实现，background_element 的 shape 设置为椭圆（oval），示例如下。

```xml
<Button
    ohos:width = "150vp"
```

```xml
    ohos:height = "50vp"
    ohos:text_size = "27fp"
    ohos:text = "button"
    ohos:background_element = "$graphic:oval_button_element"
    ohos:left_margin = "15vp"
    ohos:bottom_margin = "15vp"
    ohos:right_padding = "8vp"
    ohos:left_padding = "8vp"
    ohos:element_left = "$media:ic_btn_reload"
/>
```

graphic 目录下的 oval_button_element.xml 文件示例如下。

```xml
<?xml version = "1.0" encoding = "utf-8"?>
<shape xmlns:ohos = "http://schemas.huawei.com/res/ohos"
    ohos:shape = "oval">
    <solid
        ohos:color = "#007CFD"/>
</shape>
```

3）胶囊按钮

胶囊按钮是一种常见的按钮，设置按钮背景时将其设置为矩形形状，并且设置 ShapeElement 的 radius 的半径，示例如下。

```xml
<Button
    ohos:id = "$ + id:button"
    ohos:width = "match_content"
    ohos:height = "match_content"
    ohos:text_size = "27fp"
    ohos:text = "button"
    ohos:background_element = "$graphic:capsule_button_element"
    ohos:left_margin = "15vp"
    ohos:bottom_margin = "15vp"
    ohos:right_padding = "15vp"
    ohos:left_padding = "15vp"
/>
```

graphic 目录下的 capsule_button_element.xml 文件示例如下。

```xml
<?xml version = "1.0" encoding = "utf-8"?>
<shape xmlns:ohos = "http://schemas.huawei.com/res/ohos"
    ohos:shape = "rectangle">
    <corners
        ohos:radius = "100"/>
    <solid
        ohos:color = "#007CFD"/>
</shape>
```

4）圆形按钮

圆形按钮和椭圆按钮的区别在于组件本身的宽度和高度相同,示例如下。

```
< Button
    ohos:id = " $ + id:button"
    ohos:width = "50vp"
    ohos:height = "50vp"
    ohos:text_size = "27fp"
    ohos:background_element = " $graphic:circle_button_element"
    ohos:text = " + "
    ohos:left_margin = "15vp"
    ohos:bottom_margin = "15vp"
    ohos:right_padding = "15vp"
    ohos:left_padding = "15vp"
/>
```

graphic 目录下的 circle_button_element.xml 文件示例如下。

```
<?xml version = "1.0" encoding = "utf – 8"?>
< shape xmlns:ohos = "http://schemas.huawei.com/res/ohos"
    ohos:shape = "oval">
    < solid
        ohos:color = " #007CFD"/>
</shape>
```

利用圆形按钮、胶囊按钮、文本组件可以绘制出拨号盘的 UI,如图 4-7 所示。

相关代码请扫描二维码获取。

注：其他组件开发(TextField、Image、TabList 和 Tab、Picker、DatePicker、TimePicker、Switch、RadioButton、RadioContainer、Checkbox、ProgressBar、RoundProgressBar、ToastDialog、PopupDialog、CommonDialog、ScrollView、ListContainer、PageSlider、PageSliderIndicator、WebView)请扫描二维码获取。

图 4-7　界面效果

4.4　常用布局开发

常用布局开发主要包括 DirectionalLayout、DependentLayout、StackLayout、TableLayout、PositionLayout 和 AdaptiveBoxLayout。

4.4.1 DirectionalLayout

DirectionalLayout 是 Java UI 中的一种重要组件布局,用于将一组组件(Component)按照水平或者垂直方向排布,能够方便对齐布局内的组件。该布局和其他布局的组合,可以实现更加丰富的方式,如图 4-8 所示。

图 4-8　DirectionalLayout 示意图

1. 支持的 XML 属性

DirectionalLayout 的共有 XML 属性继承自 Component,DirectionalLayout 的自有 XML 属性如表 4-3 所示。

表 4-3　DirectionalLayout 自有 XML 属性

属性名称	中文描述	取值	取值说明	使用案例
alignment	对齐方式	left	表示左对齐	可以设置取值项如本表中所列,也可以使用"|"进行多项组合。 ohos:alignment="top|left" ohos:alignment="left"
		top	表示顶部对齐	
		right	表示右对齐	
		bottom	表示底部对齐	
		horizontal_center	表示水平居中对齐	
		vertical_center	表示垂直居中对齐	
		center	表示居中对齐	
		start	表示靠起始端对齐	
		end	表示靠结束端对齐	
orientation	子布局排列方向	horizontal	表示水平方向布局	ohos:orientation="horizontal"
		vertical	表示垂直方向布局	ohos:orientation="vertical"
total_weight	所有子视图的权重之和	float 类型	可以直接设置浮点数值,也可以引用 float 浮点数资源	ohos:total_weight="2.5" ohos:total_weight=" $float:total_weight"

DirectionalLayout 所包含组件可支持的 XML 属性如表 4-4 所示。

表 4-4 DirectionalLayout 所包含组件可支持的 XML 属性

属性名称	中文描述	取值	取值说明	使用案例
layout_alignment	对齐方式	left	表示左对齐	可以设置取值项如本表中所列,也可以使用"|"进行多项组合 ohos:layout_alignment = "top" ohos:layout_alignment = "top|left"
		top	表示顶部对齐	
		right	表示右对齐	
		bottom	表示底部对齐	
		horizontal_center	表示水平居中对齐	
		vertical_center	表示垂直居中对齐	
		center	表示居中对齐	
weight	比重	float 类型	可以直接设置浮点数值,也可以引用 float 浮点数资源	ohos:weight = "1" ohos:weight = "$float:weight"

2．排列方式

DirectionalLayout 的排列方向(orientation)分为水平(horizontal)或者垂直(vertical)。使用排列方向设置布局内组件的排列方式,默认为垂直排列。

1) 垂直排列

垂直方向排列三个按钮,相关代码如下。

```xml
<?xml version = "1.0" encoding = "utf-8"?>
<DirectionalLayout
    xmlns:ohos = "http://schemas.huawei.com/res/ohos"
    ohos:width = "match_parent"
    ohos:height = "match_content"
    ohos:orientation = "vertical">
    <Button
        ohos:width = "33vp"
        ohos:height = "20vp"
        ohos:bottom_margin = "3vp"
        ohos:left_margin = "13vp"
        ohos:background_element = " $graphic:color_cyan_element"
        ohos:text = "Button 1"/>
    <Button
        ohos:width = "33vp"
        ohos:height = "20vp"
        ohos:bottom_margin = "3vp"
        ohos:left_margin = "13vp"
        ohos:background_element = " $graphic:color_cyan_element"
        ohos:text = "Button 2"/>
    <Button
        ohos:width = "33vp"
        ohos:height = "20vp"
        ohos:bottom_margin = "3vp"
        ohos:left_margin = "13vp"
```

```xml
            ohos:background_element = " $graphic:color_cyan_element"
            ohos:text = "Button 3"/>
</DirectionalLayout>
```
color_cyan_element.xml
```xml
<?xml version = "1.0" encoding = "utf-8"?>
<shape xmlns:ohos = "http://schemas.huawei.com/res/ohos"
    ohos:shape = "rectangle">
    <solid
        ohos:color = "#00FFFD"/>
</shape>
```

2）水平排列

本部分包括水平排列超过和未超过布局本身大小的相关代码。

子组件未超过布局本身大小，水平方向排列三个按钮，相关代码如下。

```xml
<?xml version = "1.0" encoding = "utf-8"?>
<DirectionalLayout
    xmlns:ohos = "http://schemas.huawei.com/res/ohos"
    ohos:width = "match_parent"
    ohos:height = "match_content"
    ohos:orientation = "horizontal">
    <Button
        ohos:width = "33vp"
        ohos:height = "20vp"
        ohos:left_margin = "13vp"
        ohos:background_element = " $graphic:color_cyan_element"
        ohos:text = "Button 1"/>
    <Button
        ohos:width = "33vp"
        ohos:height = "20vp"
        ohos:left_margin = "13vp"
        ohos:background_element = " $graphic:color_cyan_element"
        ohos:text = "Button 2"/>
    <Button
        ohos:width = "33vp"
        ohos:height = "20vp"
        ohos:left_margin = "13vp"
        ohos:background_element = " $graphic:color_cyan_element"
        ohos:text = "Button 3"/>
</DirectionalLayout>
```
color_cyan_element.xml
```xml
<?xml version = "1.0" encoding = "utf-8"?>
<shape xmlns:ohos = "http://schemas.huawei.com/res/ohos"
    ohos:shape = "rectangle">
    <solid
        ohos:color = "#00FFFD"/>
</shape>
```

DirectionalLayout 不会自动换行，其子组件会按照设定的方向依次排列，若超过布局本身的大小，超出部分将不会被显示，相关代码如下。

```xml
<?xml version="1.0" encoding="utf-8"?>
<DirectionalLayout
    xmlns:ohos="http://schemas.huawei.com/res/ohos"
    ohos:width="match_parent"
    ohos:height="20vp"
    ohos:orientation="horizontal">
    <Button
        ohos:width="166vp"
        ohos:height="match_content"
        ohos:left_margin="13vp"
        ohos:background_element="$graphic:color_cyan_element"
        ohos:text="Button 1"/>
    <Button
        ohos:width="166vp"
        ohos:height="match_content"
        ohos:left_margin="13vp"
        ohos:background_element="$graphic:color_cyan_element"
        ohos:text="Button 2"/>
    <Button
        ohos:width="166vp"
        ohos:height="match_content"
        ohos:left_margin="13vp"
        ohos:background_element="$graphic:color_cyan_element"
        ohos:text="Button 3"/>
</DirectionalLayout>
```

color_cyan_element.xml：

```xml
<?xml version="1.0" encoding="utf-8"?>
<shape xmlns:ohos="http://schemas.huawei.com/res/ohos"
    ohos:shape="rectangle">
    <solid
        ohos:color="#00FFFD"/>
</shape>
```

3．对齐方式

DirectionalLayout 中的组件使用 layout_alignment 控制自身在布局中的对齐方式。对齐方式和排列方式密切相关，当排列方式为水平方向时，可选的对齐方式只有作用于垂直方向的类型（top、bottom、vertical_center、center），其他对齐方式不会生效。当排列方式为垂直方向时，可选的对齐方式只有作用于水平方向的类型（left、right、start、end、horizontal_center、center），其他对齐方式不会生效，左、中、右三种对齐方式代码如下。

```xml
<?xml version="1.0" encoding="utf-8"?>
<DirectionalLayout
    xmlns:ohos="http://schemas.huawei.com/res/ohos"
```

```xml
        ohos:width = "match_parent"
        ohos:height = "60vp">
    <Button
            ohos:width = "50vp"
            ohos:height = "20vp"
            ohos:background_element = "$graphic:color_cyan_element"
            ohos:layout_alignment = "left"
            ohos:text = "Button 1"/>
    <Button
            ohos:width = "50vp"
            ohos:height = "20vp"
            ohos:background_element = "$graphic:color_cyan_element"
            ohos:layout_alignment = "horizontal_center"
            ohos:text = "Button 2"/>
    <Button
            ohos:width = "50vp"
            ohos:height = "20vp"
            ohos:background_element = "$graphic:color_cyan_element"
            ohos:layout_alignment = "right"
            ohos:text = "Button 3"/>
</DirectionalLayout>
```
color_cyan_element.xml:
```xml
<?xml version = "1.0" encoding = "utf-8"?>
<shape xmlns:ohos = "http://schemas.huawei.com/res/ohos"
    ohos:shape = "rectangle">
    <solid
        ohos:color = "#00FFFD"/>
</shape>
```

4. 权重

权重(weight)是按比例分配组件占用父组件的大小,在水平布局下计算公式为:父布局可分配宽度＝父布局宽度－所有子组件 width 之和;组件宽度＝组件 weight/所有组件 weight 之和×父布局可分配宽度。建议使用 width＝0 按比例分配父布局的宽度。

1) 三个组件的 1∶1∶1 效果代码

```xml
<?xml version = "1.0" encoding = "utf-8"?>
<DirectionalLayout
        xmlns:ohos = "http://schemas.huawei.com/res/ohos"
        ohos:width = "match_parent"
        ohos:height = "match_content"
        ohos:orientation = "horizontal">
    <Button
            ohos:width = "0vp"
            ohos:height = "20vp"
            ohos:weight = "1"
```

```
                ohos:background_element = " $graphic:color_cyan_element"
                ohos:text = "Button 1"/>
        < Button
                ohos:width = "0vp"
                ohos:height = "20vp"
                ohos:weight = "1"
                ohos:background_element = " $graphic:color_gray_element"
                ohos:text = "Button 2"/>
        < Button
                ohos:width = "0vp"
                ohos:height = "20vp"
                ohos:weight = "1"
                ohos:background_element = " $graphic:color_cyan_element"
                ohos:text = "Button 3"/>
</DirectionalLayout >
```

2) color_cyan_element.xml

```
<?xml version = "1.0" encoding = "utf-8"?>
< shape xmlns:ohos = "http://schemas.huawei.com/res/ohos"
        ohos:shape = "rectangle">
        < solid
                ohos:color = " #00FFFD"/>
</shape >
```

3) color_gray_element.xml

```
<?xml version = "1.0" encoding = "utf-8"?>
< shape xmlns:ohos = "http://schemas.huawei.com/res/ohos"
        ohos:shape = "rectangle">
        < solid
                ohos:color = " #878787"/>
</shape >
```

4.4.2 DependentLayout

DependentLayout 是 Java UI 框架里的一种常见布局。与 DirectionalLayout 相比，DependentLayout 拥有更多的排布方式，每个组件可以指定相对于其他同级元素的位置，或者指定相对于父组件的位置，如图 4-9 所示。

1. 支持的 XML 属性

DependentLayout 的共有 XML 属性继承自 Component，如表 4-5 所示。

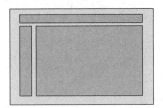

图 4-9 DependentLayout 示意图

表 4-5 DependentLayout 的共有 XML 属性

属性名称	中文描述	取值	取值说明	使用案例
alignment	对齐方式	left	表示左对齐	可以设置取值项如本表中所列,也可以使用"|"进行多项组合 ohos:alignment="top|left" ohos:alignment="left"
		top	表示顶部对齐	
		right	表示右对齐	
		bottom	表示底部对齐	
		horizontal_center	表示水平居中对齐	
		vertical_center	表示垂直居中对齐	
		center	表示居中对齐	

DependentLayout 所包含组件可支持的 XML 属性请扫描二维码获取。

2. 排列方式

DependentLayout 的排列方式是相对于其他同级组件或者父组件的位置进行布局。

1) 相对于同级组件的对齐

相对于同级组件的对齐方式如下。

(1) 根据位置对齐。left_of、right_of、start_of、end_of、above 和 below 均相对于同级组件的不同位置进行对齐。例如,left_of 对齐方式是将右边缘与同级组件的左边缘对齐,对齐后位于同级组件的左侧。其他几种对齐方式遵循的逻辑与此相同,需要注意的是,start_of 和 end_of 会跟随当前布局起始方向变化,如图 4-10 所示。

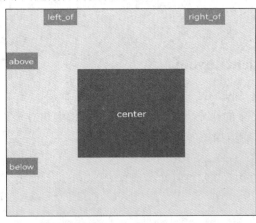

图 4-10 位置对齐方式

相关代码如下。

```xml
<?xml version = "1.0" encoding = "utf-8"?>
<DependentLayout
    xmlns:ohos = "http://schemas.huawei.com/res/ohos"
    ohos:height = "500vp"
    ohos:width = "500vp"
    ohos:background_element = "#EDEDED">
    <Text
        ohos:id = "$+id:text_center"
        ohos:height = "210vp"
        ohos:width = "210vp"
        ohos:background_element = "#878787"
        ohos:center_in_parent = "true"
        ohos:text = "center"
        ohos:text_alignment = "center"
        ohos:text_size = "20fp"
        ohos:text_color = "#FFFFFF"/>
    <Text
        ohos:id = "$+id:text_left"
        ohos:height = "match_content"
        ohos:width = "match_content"
        ohos:left_of = "$id:text_center"
        ohos:background_element = "#FF9912"
        ohos:padding = "8vp"
        ohos:text = "left_of"
        ohos:text_size = "18fp"
        ohos:text_color = "#FFFFFF"/>
    <Text
        ohos:id = "$+id:text_right"
        ohos:height = "match_content"
        ohos:width = "match_content"
        ohos:right_of = "$id:text_center"
        ohos:background_element = "#FF9912"
        ohos:padding = "8vp"
        ohos:text = "right_of"
        ohos:text_size = "18fp"
        ohos:text_color = "#FFFFFF"/>
    <Text
        ohos:id = "$+id:text_above"
        ohos:height = "match_content"
        ohos:width = "match_content"
        ohos:above = "$id:text_center"
        ohos:background_element = "#FF9912"
        ohos:padding = "8vp"
        ohos:text = "above"
        ohos:text_size = "18fp"
```

```
        ohos:text_color = "#FFFFFF"/>
    <Text
        ohos:id = "$+id:text_below"
        ohos:height = "match_content"
        ohos:width = "match_content"
        ohos:below = "$id:text_center"
        ohos:background_element = "#FF9912"
        ohos:padding = "8vp"
        ohos:text = "below"
        ohos:text_size = "18fp"
        ohos:text_color = "#FFFFFF"/>
</DependentLayout>
```

（2）根据边对齐。align_left、align_right、align_top、align_bottom、align_start 和 align_end 都是与同级组件的相同边对齐。例如，align_left 对齐方式是将当前组件与同级组件的左边缘对齐。其他几种对齐方式遵循的逻辑与此相同，需要注意的是，align_start 和 align_end 会跟随当前布局起始方向变化，如图 4-11 所示。

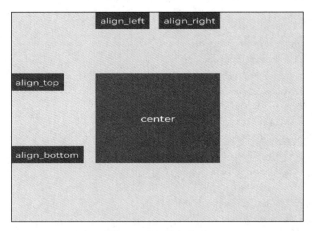

图 4-11　边对齐方式

相关代码如下。

```
<?xml version = "1.0" encoding = "utf-8"?>
<DependentLayout
    xmlns:ohos = "http://schemas.huawei.com/res/ohos"
    ohos:height = "500vp"
    ohos:width = "500vp"
    ohos:background_element = "#EDEDED">
    <Text
        ohos:id = "$+id:text_center"
        ohos:height = "210vp"
        ohos:width = "210vp"
        ohos:background_element = "#878787"
```

```xml
    ohos:center_in_parent = "true"
    ohos:text = "center"
    ohos:text_alignment = "center"
    ohos:text_size = "20fp"
    ohos:text_color = "#FFFFFF"/>
<Text
    ohos:id = "$+id:text_align_top"
    ohos:height = "match_content"
    ohos:width = "match_content"
    ohos:align_top = "$id:text_center"
    ohos:background_element = "#228B22"
    ohos:padding = "8vp"
    ohos:text = "align_top"
    ohos:text_size = "18fp"
    ohos:text_color = "#FFFFFF"/>
<Text
    ohos:id = "$+id:text_align_bottom"
    ohos:height = "match_content"
    ohos:width = "match_content"
    ohos:align_bottom = "$id:text_center"
    ohos:background_element = "#228B22"
    ohos:padding = "8vp"
    ohos:text = "align_bottom"
    ohos:text_size = "18fp"
    ohos:text_color = "#FFFFFF"/>
<Text
    ohos:id = "$+id:text_align_left"
    ohos:height = "match_content"
    ohos:width = "match_content"
    ohos:align_left = "$id:text_center"
    ohos:background_element = "#228B22"
    ohos:padding = "8vp"
    ohos:text = "align_left"
    ohos:text_size = "18fp"
    ohos:text_color = "#FFFFFF"/>
<Text
    ohos:id = "$+id:text_align_right"
    ohos:height = "match_content"
    ohos:width = "match_content"
    ohos:align_right = "$id:text_center"
    ohos:background_element = "#228B22"
    ohos:padding = "8vp"
    ohos:text = "align_right"
    ohos:text_size = "18fp"
    ohos:text_color = "#FFFFFF"/>
</DependentLayout>
```

（3）组合属性对齐方式。在逻辑不冲突的情况下，开发者可以组合使用多种对齐方式进行布局，如图 4-12 所示，相关代码扫描二维码获取。

2）相对于父级组件的对齐

相对于父级组件的对齐包括单个属性对齐方式和组合属性对齐方式。

（1）单个属性对齐方式如图 4-13 所示。

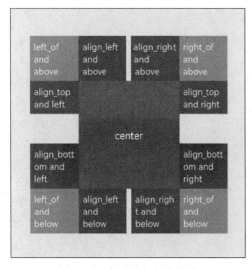

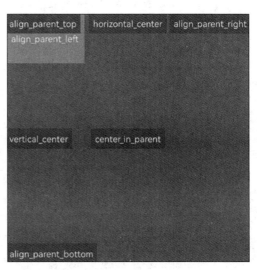

图 4-12　常用对齐方式效果　　　　　图 4-13　单个属性对齐方式

相关代码如下。

```xml
<?xml version = "1.0" encoding = "utf-8"?>
<DependentLayout
    xmlns:ohos = "http://schemas.huawei.com/res/ohos"
    ohos:height = "500vp"
    ohos:width = "500vp"
    ohos:background_element = "#878787">
    <Text
        ohos:height = "100vp"
        ohos:width = "match_content"
        ohos:align_parent_left = "true"
        ohos:background_element = "#FF9912"
        ohos:padding = "12vp"
        ohos:multiple_lines = "true"
        ohos:text = "align_parent_left"
        ohos:text_alignment = "vertical_center"
        ohos:text_size = "18fp"
        ohos:text_color = "#FFFFFF"/>
    <Text
        ohos:height = "match_content"
```

```
        ohos:width = "match_content"
        ohos:align_parent_right = "true"
        ohos:background_element = "#228B22"
        ohos:padding = "8vp"
        ohos:multiple_lines = "true"
        ohos:text = "align_parent_right"
        ohos:text_size = "18fp"
        ohos:text_color = "#FFFFFF"/>
    <Text
        ohos:height = "match_content"
        ohos:width = "match_content"
        ohos:align_parent_top = "true"
        ohos:background_element = "#228B22"
        ohos:padding = "8vp"
        ohos:multiple_lines = "true"
        ohos:text = "align_parent_top"
        ohos:text_size = "18fp"
        ohos:text_color = "#FFFFFF"/>
    <Text
        ohos:height = "match_content"
        ohos:width = "match_content"
        ohos:align_parent_bottom = "true"
        ohos:background_element = "#228B22"
        ohos:padding = "8vp"
        ohos:multiple_lines = "true"
        ohos:text = "align_parent_bottom"
        ohos:text_size = "18fp"
        ohos:text_color = "#FFFFFF"/>
    <Text
        ohos:height = "match_content"
        ohos:width = "match_content"
        ohos:center_in_parent = "true"
        ohos:background_element = "#228B22"
        ohos:padding = "8vp"
        ohos:multiple_lines = "true"
        ohos:text = "center_in_parent"
        ohos:text_size = "18fp"
        ohos:text_color = "#FFFFFF"/>
    <Text
        ohos:height = "match_content"
        ohos:width = "match_content"
        ohos:horizontal_center = "true"
        ohos:background_element = "#228B22"
        ohos:padding = "8vp"
        ohos:multiple_lines = "true"
        ohos:text = "horizontal_center"
        ohos:text_size = "18fp"
```

```
            ohos:text_color = "#FFFFFF"/>
        <Text
            ohos:height = "match_content"
            ohos:width = "match_content"
            ohos:vertical_center = "true"
            ohos:background_element = "#228B22"
            ohos:padding = "8vp"
            ohos:multiple_lines = "true"
            ohos:text = "vertical_center"
            ohos:text_size = "18fp"
            ohos:text_color = "#FFFFFF"/>
</DependentLayout>
```

（2）组合属性对齐方式如图 4-14 所示。

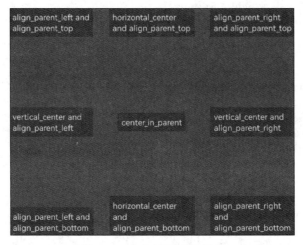

图 4-14　组合属性对齐方式

相关代码如下。

```
<?xml version = "1.0" encoding = "utf-8"?>
<DependentLayout
    xmlns:ohos = "http://schemas.huawei.com/res/ohos"
    ohos:height = "500vp"
    ohos:width = "640vp"
    ohos:background_element = "#878787">
    <Text
        ohos:height = "match_content"
        ohos:width = "match_content"
        ohos:center_in_parent = "true"
        ohos:background_element = "#228B22"
        ohos:padding = "8vp"
        ohos:multiple_lines = "true"
        ohos:text = "center_in_parent"
```

```
        ohos:text_size = "18fp"
        ohos:text_color = "#FFFFFF"/>
<Text
    ohos:height = "match_content"
    ohos:width = "190vp"
    ohos:align_parent_left = "true"
    ohos:align_parent_top = "true"
    ohos:background_element = "#228B22"
    ohos:padding = "8vp"
    ohos:multiple_lines = "true"
    ohos:text = "align_parent_left and align_parent_top"
    ohos:text_size = "18fp"
    ohos:text_color = "#FFFFFF"/>
<Text
    ohos:height = "match_content"
    ohos:width = "190vp"
    ohos:align_parent_left = "true"
    ohos:align_parent_bottom = "true"
    ohos:background_element = "#228B22"
    ohos:padding = "8vp"
    ohos:multiple_lines = "true"
    ohos:text = "align_parent_left and align_parent_bottom"
    ohos:text_size = "18fp"
    ohos:text_color = "#FFFFFF"/>
<Text
    ohos:height = "match_content"
    ohos:width = "190vp"
    ohos:align_parent_right = "true"
    ohos:align_parent_top = "true"
    ohos:background_element = "#228B22"
    ohos:padding = "8vp"
    ohos:multiple_lines = "true"
    ohos:text = "align_parent_right and align_parent_top"
    ohos:text_size = "18fp"
    ohos:text_color = "#FFFFFF"/>
<Text
    ohos:height = "match_content"
    ohos:width = "190vp"
    ohos:align_parent_right = "true"
    ohos:align_parent_bottom = "true"
    ohos:background_element = "#228B22"
    ohos:padding = "8vp"
    ohos:multiple_lines = "true"
    ohos:text = "align_parent_right and align_parent_bottom"
    ohos:text_size = "18fp"
    ohos:text_color = "#FFFFFF"/>
<Text
```

```
        ohos:height = "match_content"
        ohos:width = "190vp"
        ohos:horizontal_center = "true"
        ohos:align_parent_top = "true"
        ohos:background_element = "#228B22"
        ohos:padding = "8vp"
        ohos:multiple_lines = "true"
        ohos:text = "horizontal_center and align_parent_top"
        ohos:text_size = "18fp"
        ohos:text_color = "#FFFFFF"/>
    <Text
        ohos:height = "match_content"
        ohos:width = "190vp"
        ohos:horizontal_center = "true"
        ohos:align_parent_bottom = "true"
        ohos:background_element = "#228B22"
        ohos:padding = "8vp"
        ohos:multiple_lines = "true"
        ohos:text = "horizontal_center and align_parent_bottom"
        ohos:text_size = "18fp"
        ohos:text_color = "#FFFFFF"/>
    <Text
        ohos:height = "match_content"
        ohos:width = "190vp"
        ohos:vertical_center = "true"
        ohos:align_parent_left = "true"
        ohos:background_element = "#228B22"
        ohos:padding = "8vp"
        ohos:multiple_lines = "true"
        ohos:text = "vertical_center and align_parent_left"
        ohos:text_size = "18fp"
        ohos:text_color = "#FFFFFF"/>
    <Text
        ohos:height = "match_content"
        ohos:width = "190vp"
        ohos:vertical_center = "true"
        ohos:align_parent_right = "true"
        ohos:background_element = "#228B22"
        ohos:padding = "8vp"
        ohos:multiple_lines = "true"
        ohos:text = "vertical_center and align_parent_right"
        ohos:text_size = "18fp"
        ohos:text_color = "#FFFFFF"/>
</DependentLayout>
```

4.4.3　StackLayout

StackLayout 直接在屏幕上开辟出一块空白的区域,添加到布局中的视图以层叠的方式显示,而它会把这些视图默认放到此区域的左上角,第一个添加到布局中的视图显示在底层,最后一个被放在顶层。上一层的视图会覆盖下一层的视图,如图 4-15 所示。

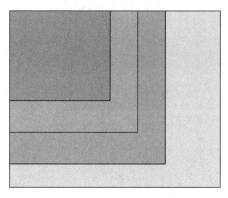

图 4-15　StackLayout 示意图

1. 支持的 XML 属性

StackLayout 无自有的 XML 属性,共有 XML 属性继承自 Component,StackLayout 所包含组件可支持的 XML 属性如表 4-6 所示。

表 4-6　StackLayout 所包含组件可支持的 XML 属性

属性名称	中文描述	取值	取值说明	使用案例
layout_alignment	对齐方式	left	表示左对齐	可以设置取值项如本表中所列,也可以使用"|"进行多项组合 ohos:layout_alignment="top" ohos:layout_alignment="top|left"
		top	表示顶部对齐	
		right	表示右对齐	
		bottom	表示底部对齐	
		horizontal_center	表示水平居中对齐	
		vertical_center	表示垂直居中对齐	
		center	表示居中对齐	

2. 创建和使用 StackLayout

StackLayout 的创建和使用方法如下。

1) 创建 StackLayout

```
<?xml version = "1.0" encoding = "utf-8"?>
<StackLayout
    xmlns:ohos = "http://schemas.huawei.com/res/ohos"
    ohos:height = "match_parent"
    ohos:width = "match_parent">
</StackLayout>
```

2) 使用默认布局添加组件

StackLayout 中组件的布局默认在区域的左上角,并且以后创建的组件会在上层,如

图 4-16 所示。

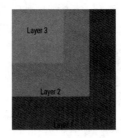

图 4-16　多个视图排列效果

XML 布局如下。

```xml
<?xml version = "1.0" encoding = "utf-8"?>
<StackLayout
    xmlns:ohos = "http://schemas.huawei.com/res/ohos"
    ohos:id = "$+id:stack_layout"
    ohos:height = "match_parent"
    ohos:width = "match_parent">
    <Text
        ohos:id = "$+id:text_blue"
        ohos:text_alignment = "bottom|horizontal_center"
        ohos:text_size = "24fp"
        ohos:text = "Layer 1"
        ohos:height = "400vp"
        ohos:width = "400vp"
        ohos:background_element = "#3F56EA" />
    <Text
        ohos:id = "$+id:text_light_purple"
        ohos:text_alignment = "bottom|horizontal_center"
        ohos:text_size = "24fp"
        ohos:text = "Layer 2"
        ohos:height = "300vp"
        ohos:width = "300vp"
        ohos:background_element = "#00AAEE" />
    <Text
        ohos:id = "$+id:text_orange"
        ohos:text_alignment = "center"
        ohos:text_size = "24fp"
        ohos:text = "Layer 3"
        ohos:height = "80vp"
        ohos:width = "80vp"
        ohos:background_element = "#00BFC9" />
</StackLayout>
```

3）使用相对位置添加组件

使用 layout_alignment 属性可以指定组件在 StackLayout 中的相对位置，Button 组件位于 StackLayout 右侧的 XML 布局代码如下。

```xml
<?xml version = "1.0" encoding = "utf-8"?>
<StackLayout
    xmlns:ohos = "http://schemas.huawei.com/res/ohos"
    ohos:id = "$+id:stack_layout"
    ohos:height = "match_parent"
    ohos:width = "match_parent">
    <Button
        ohos:id = "$+id:button"
        ohos:height = "40vp"
        ohos:width = "80vp"
        ohos:layout_alignment = "right"
        ohos:background_element = "#3399FF"/>
</StackLayout>
```

4.4.4 TableLayout

TableLayout 使用表格的方式划分子组件，如图 4-17 所示。

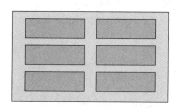

图 4-17 表格布局示意图

1. 支持的 XML 属性

TableLayout 的共有 XML 属性继承自 Component，如表 4-7 所示。

表 4-7 TableLayout 的自有 XML 属性

属性名称	中文描述	取值	取值说明	使用案例
alignment_type	对齐方式	align_edges	表示 TableLayout 内的组件按边界对齐	ohos:alignment_type="align_edges"
		align_contents	表示 TableLayout 内的组件按边距对齐	ohos:alignment_type="align_contents"
column_count	列数	integer 类型	既可以直接设置整型数值，也可以引用 integer 资源	ohos:column_count="3" ohos:column_count="$integer:count"

续表

属性名称	中文描述	取值	取值说明	使用案例
row_count	行数	integer 类型	既可以直接设置整型数值,也可以引用 integer 资源	ohos:row_count="2" ohos:row_count="$integer:count"
orientation	排列方向	horizontal	表示水平方向布局	ohos:orientation="horizontal"
		vertical	表示垂直方向布局	ohos:orientation="vertical"

2. 创建 TableLayout

TableLayout 的创建方法如下。

1) 在 XML 中创建 TableLayout

相关代码如下。

```xml
<?xml version="1.0" encoding="utf-8"?>
<TableLayout
    xmlns:ohos="http://schemas.huawei.com/res/ohos"
    ohos:height="match_parent"
    ohos:width="match_parent"
    ohos:background_element="#87CEEB"
    ohos:padding="8vp">
</TableLayout>
```

2) 添加子组件

添加子组件方法如下。

(1) 在 graphic 文件夹下创建 Text 的背景 table_text_bg_element.xml,相关代码如下。

```xml
<?xml version="1.0" encoding="utf-8"?>
<shape xmlns:ohos="http://schemas.huawei.com/res/ohos"
       ohos:shape="rectangle">
    <corners
        ohos:radius="5vp"/>
    <stroke
        ohos:width="1vp"
        ohos:color="gray"/>
    <solid
        ohos:color="#00BFFF"/>
</shape>
```

(2) 在 TableLayout 布局中添加子组件,相关代码如下。

```xml
<?xml version="1.0" encoding="utf-8"?>
<TableLayout
    xmlns:ohos="http://schemas.huawei.com/res/ohos"
```

```
        ohos:height = "match_parent"
        ohos:width = "match_parent"
        ohos:background_element = "#87CEEB"
        ohos:padding = "8vp">
    <Text
        ohos:height = "60vp"
        ohos:width = "60vp"
        ohos:background_element = "$graphic:table_text_bg_element"
        ohos:margin = "8vp"
        ohos:text = "1"
        ohos:text_alignment = "center"
        ohos:text_size = "20fp"/>
    <Text
        ohos:height = "60vp"
        ohos:width = "60vp"
        ohos:background_element = "$graphic:table_text_bg_element"
        ohos:margin = "8vp"
        ohos:text = "2"
        ohos:text_alignment = "center"
        ohos:text_size = "20fp"/>
    <Text
        ohos:height = "60vp"
        ohos:width = "60vp"
        ohos:background_element = "$graphic:table_text_bg_element"
        ohos:margin = "8vp"
        ohos:text = "3"
        ohos:text_alignment = "center"
        ohos:text_size = "20fp"/>
    <Text
        ohos:height = "60vp"
        ohos:width = "60vp"
        ohos:background_element = "$graphic:table_text_bg_element"
        ohos:margin = "8vp"
        ohos:text = "4"
        ohos:text_alignment = "center"
        ohos:text_size = "20fp"/>
</TableLayout>
```

3. 设置行列数

TableLayout 默认一列多行，设置 TableLayout 的行为 2，列为 2，相关代码如下。

```
<TableLayout
    ...
    ohos:row_count = "2"
    ohos:column_count = "2">
```

4. 设置布局排列方向

在 XML 中设置布局排列方向，以 vertical 为例，按照垂直顺序排列，相关代码如下。

```xml
<TableLayout
    ...
    ohos:orientation="vertical">
    ...
</TableLayout>
```

5. 设置对齐方式

TableLayout 提供两种对齐方式,边距对齐 align_contents、边界对齐 align_edges,默认为边距对齐 align_contents。

(1) 边距对齐方式,相关代码如下。

```xml
<?xml version="1.0" encoding="utf-8"?>
<TableLayout
    xmlns:ohos="http://schemas.huawei.com/res/ohos"
    ohos:height="match_content"
    ohos:width="match_content"
    ohos:alignment_type="align_contents"
    ohos:background_element="$graphic:layout_borderline"
    ohos:column_count="3"
    ohos:padding="8vp">
    <Text
        ohos:height="48vp"
        ohos:width="48vp"
        ohos:background_element="$graphic:table_text_bg_element"
        ohos:margin="8vp"
        ohos:padding="8vp"
        ohos:text="1"
        ohos:text_alignment="center"
        ohos:text_size="14fp"/>
    <Text
        ohos:height="48vp"
        ohos:width="48vp"
        ohos:background_element="$graphic:table_text_bg_element"
        ohos:margin="16vp"
        ohos:padding="8vp"
        ohos:text="2"
        ohos:text_alignment="center"
        ohos:text_size="14fp"/>
    <Text
        ohos:height="48vp"
        ohos:width="48vp"
        ohos:background_element="$graphic:table_text_bg_element"
        ohos:margin="32vp"
        ohos:padding="8vp"
        ohos:text="3"
        ohos:text_alignment="center"
```

```
            ohos:text_size = "14fp"/>
        <Text
            ohos:height = "48vp"
            ohos:width = "48vp"
            ohos:background_element = " $graphic:table_text_bg_element"
            ohos:margin = "32vp"
            ohos:padding = "8vp"
            ohos:text = "4"
            ohos:text_alignment = "center"
            ohos:text_size = "14fp"/>
        <Text
            ohos:height = "48vp"
            ohos:width = "48vp"
            ohos:background_element = " $graphic:table_text_bg_element"
            ohos:margin = "16vp"
            ohos:padding = "8vp"
            ohos:text = "5"
            ohos:text_alignment = "center"
            ohos:text_size = "14fp"/>
        <Text
            ohos:height = "48vp"
            ohos:width = "48vp"
            ohos:background_element = " $graphic:table_text_bg_element"
            ohos:margin = "8vp"
            ohos:padding = "8vp"
            ohos:text = "6"
            ohos:text_alignment = "center"
            ohos:text_size = "14fp"/>
</TableLayout>
```

（2）边界对齐方式，将 TableLayout 的对齐方式修改为边界对齐，相关代码如下。

```
<TableLayout
    ...
    ohos:alignment_type = "align_edges">
    ...
</TableLayout>
```

（3）引用 graphic 文件夹下的背景资源文件为 layout_borderline.xml，相关代码如下。

```
<?xml version = "1.0" encoding = "utf-8"?>
<shape xmlns:ohos = "http://schemas.huawei.com/res/ohos"
       ohos:shape = "rectangle">
    <corners
        ohos:radius = "5vp"/>
    <stroke
        ohos:width = "1vp"
        ohos:color = "gray"/>
```

```
</shape>
```

6. 设置子组件的行列属性

本部分包括设置子组件的行列属性和设置子组件的权重。

1) 实现合并单元格的效果

TableLayout 合并单元格的效果可以通过设置子组件的行列属性实现,如图 4-18 所示。

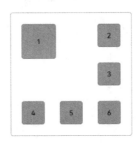

图 4-18　设置子组件的行列属性均为 2 的效果展示

(1) 在 XML 中创建 TableLayout,并添加子组件,相关代码如下。

```
<?xml version = "1.0" encoding = "utf - 8"?>
< TableLayout
    xmlns:ohos = "http://schemas.huawei.com/res/ohos"
    ohos:height = "match_content"
    ohos:width = "match_content"
    ohos:alignment_type = "align_edges"
    ohos:background_element = " $graphic:layout_borderline"
    ohos:column_count = "3"
    ohos:padding = "8vp"
    ohos:row_count = "3">
    < Text
        ohos:id = " $ + id:text_one"
        ohos:height = "48vp"
        ohos:width = "48vp"
        ohos:background_element = " $graphic:table_text_bg_element"
        ohos:margin = "16vp"
        ohos:padding = "8vp"
        ohos:text = "1"
        ohos:text_alignment = "center"
        ohos:text_size = "14fp"/>
    < Text
        ohos:height = "48vp"
        ohos:width = "48vp"
        ohos:background_element = " $graphic:table_text_bg_element"
        ohos:margin = "16vp"
        ohos:padding = "8vp"
        ohos:text = "2"
```

```
            ohos:text_alignment = "center"
            ohos:text_size = "14fp"/>
    < Text
        ohos:height = "48vp"
        ohos:width = "48vp"
        ohos:background_element = " $graphic:table_text_bg_element"
        ohos:margin = "16vp"
        ohos:padding = "8vp"
        ohos:text = "3"
        ohos:text_alignment = "center"
        ohos:text_size = "14fp"/>
    < Text
        ohos:height = "48vp"
        ohos:width = "48vp"
        ohos:background_element = " $graphic:table_text_bg_element"
        ohos:margin = "16vp"
        ohos:padding = "8vp"
        ohos:text = "4"
        ohos:text_alignment = "center"
        ohos:text_size = "14fp"/>
    < Text
        ohos:height = "48vp"
        ohos:width = "48vp"
        ohos:background_element = " $graphic:table_text_bg_element"
        ohos:margin = "16vp"
        ohos:padding = "8vp"
        ohos:text = "5"
        ohos:text_alignment = "center"
        ohos:text_size = "14fp"/>
    < Text
        ohos:height = "48vp"
        ohos:width = "48vp"
        ohos:background_element = " $graphic:table_text_bg_element"
        ohos:margin = "16vp"
        ohos:padding = "8vp"
        ohos:text = "6"
        ohos:text_alignment = "center"
        ohos:text_size = "14fp"/>
</TableLayout >
```

（2）在Java代码中设置子组件的行列属性，相关代码如下。

```
@Override
    protected void onStart(Intent intent) {
        ...
        Component component = findComponentById(ResourceTable.Id_text_one);
        TableLayout.LayoutConfig tlc = new TableLayout.LayoutConfig(vp2px(72), vp2px(72));
```

```
            tlc.columnSpec = TableLayout.specification(TableLayout.DEFAULT, 2);
            tlc.rowSpec = TableLayout.specification(TableLayout.DEFAULT, 2);
            component.setLayoutConfig(tlc);
    }
    private int vp2px(float vp) {
            return AttrHelper.vp2px(vp, getContext());
    }
```

在设置子组件的行列属性时，TableLayout 剩余的行数和列数必须大于等于该子组件所设置的行数和列数。目前仅支持 Java 代码设置 TableLayout 子组件的行列属性。

（3）在创建子组件的行列属性时，还可设置子组件的对齐方式，如图 4-19 所示，修改上述 Java 代码如下。

```
@Override
    protected void onStart(Intent intent) {
            ...
             tlc.columnSpec = TableLayout.specification(TableLayout.DEFAULT, 2, TableLayout.Alignment.ALIGNMENT_FILL);
              tlc.rowSpec = TableLayout.specification(TableLayout.DEFAULT, 2, TableLayout.Alignment.ALIGNMENT_FILL);
            ...
    }
```

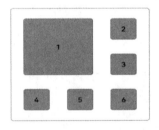

图 4-19　设置子组件对齐方式

2）设置子组件的权重

相关代码如下。

```
@Override
    protected void onStart(Intent intent) {
            ...
            TableLayout.LayoutConfig tlc = new TableLayout.LayoutConfig(0, vp2px(48));
            tlc.columnSpec = TableLayout.specification(TableLayout.DEFAULT, 1, 1.0f);
            tlc.rowSpec = TableLayout.specification(TableLayout.DEFAULT, 1);
            findComponentById(ResourceTable.Id_text_one).setLayoutConfig(tlc);
            findComponentById(ResourceTable.Id_text_two).setLayoutConfig(tlc);
            findComponentById(ResourceTable.Id_text_three).setLayoutConfig(tlc);
            findComponentById(ResourceTable.Id_text_four).setLayoutConfig(tlc);
```

```
        findComponentById(ResourceTable.Id_text_five).setLayoutConfig(tlc);
        findComponentById(ResourceTable.Id_text_six).setLayoutConfig(tlc);
    }
```

上述代码将子组件的宽度权重设置为 1.0，每行子组件会均分 TableLayout 的宽度，所以需要设置 TableLayout 为固定宽度或 match_parent，相关代码如下。

```
<TableLayout
    ohos:width = "match_parent"
    …>
    <Text
        ohos:id = " $ + id:text_one"
        …/>
    <Text
        ohos:id = " $ + id:text_two"
        …/>
    <Text
        ohos:id = " $ + id:text_three"
        …/>
    <Text
        ohos:id = " $ + id:text_four"
        …/>
    <Text
        ohos:id = " $ + id:text_five"
        …/>
    <Text
        ohos:id = " $ + id:text_six"
        …/>
</TableLayout>
```

4.4.5 PositionLayout

在 PositionLayout 中，子组件通过指定准确的 x/y 坐标值在屏幕上显示。(0,0) 为左上角，当向下或向右移动时，坐标值变大，允许组件之间互相重叠，如图 4-20 所示。

（1）布局方式。PositionLayout 以坐标的形式控制组件的显示位置，允许组件相互重叠。在 layout 目录下的 XML 文件中创建 PositionLayout 并添加多个组件，通过 position_x 和 position_y 属性设置子组件的坐标，相关代码如下。

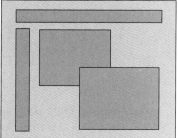

图 4-20 设定坐标值

```
<?xml version = "1.0" encoding = "utf - 8"?>
<PositionLayout
    xmlns:ohos = "http://schemas.huawei.com/res/ohos"
    ohos:id = " $ + id:position"
```

```xml
        ohos:height = "match_parent"
        ohos:width = "300vp"
        ohos:background_element = "#3387CEFA">
    <Text
        ohos:id = "$+id:position_text_1"
        ohos:height = "50vp"
        ohos:width = "200vp"
        ohos:background_element = "#9987CEFA"
        ohos:position_x = "50vp"
        ohos:position_y = "8vp"
        ohos:text = "Title"
        ohos:text_alignment = "center"
        ohos:text_size = "20fp"/>
    <Text
        ohos:id = "$+id:position_text_2"
        ohos:height = "200vp"
        ohos:width = "200vp"
        ohos:background_element = "#9987CEFA"
        ohos:position_x = "8vp"
        ohos:position_y = "64vp"
        ohos:text = "Content"
        ohos:text_alignment = "center"
        ohos:text_size = "20fp"/>
    <Text
        ohos:id = "$+id:position_text_3"
        ohos:height = "200vp"
        ohos:width = "200vp"
        ohos:background_element = "#9987CEFA"
        ohos:position_x = "92vp"
        ohos:position_y = "188vp"
        ohos:text = "Content"
        ohos:text_alignment = "center"
        ohos:text_size = "20fp"/>
</PositionLayout>
```

（2）设置子组件坐标时（position_x 和 position_y 属性），除了上述示例中的 XML 方式，还可以在对应的 AbilitySlice 中通过 setPosition(int x, int y)接口设置，Java 示例代码如下。

```java
Text title = (Text)findComponentById(ResourceTable.Id_position_text_1);
Text content1 = (Text)findComponentById(ResourceTable.Id_position_text_2);
Text content2 = (Text)findComponentById(ResourceTable.Id_position_text_3);
        title.setPosition(vp2px(50), vp2px(8));
        content1.setPosition(vp2px(8), vp2px(64));
        content2.setPosition(vp2px(92), vp2px(188));
```

（3）单位转换方法如下。

```
private int vp2px(float vp){
        return AttrHelper.vp2px(vp,this);
}
```

（4）对于超过布局本身大小的组件，超出部分不显示，示例代码如下。

```
<?xml version = "1.0" encoding = "utf - 8"?>
< PositionLayout
    …>
    …
    < Text
        ohos:id = " $ + id:position_text_4"
        ohos:height = "120vp"
        ohos:width = "120vp"
        ohos:background_element = " ♯9987CEFA"
        ohos:position_x = "212vp"
        ohos:position_y = "64vp"
        ohos:text = "Right"
        ohos:text_alignment = "center"
        ohos:text_size = "20fp"/>
</PositionLayout >
```

4.4.6 AdaptiveBoxLayout

AdaptiveBoxLayout 是自适应盒子布局，它具备在不同屏幕尺寸设备上的自适应布局能力，主要用于相同级别的多个组件需要在不同屏幕尺寸设备上自动调整列数的场景。

该布局中的每个子组件用一个单独的"盒子"装起来，子组件设置的布局参数以盒子作为父布局生效，不以整个自适应布局为生效范围。每个盒子的宽度固定为布局总宽度除以自适应得到的列数，高度为 match_content，每行中的所有盒子按最高的高度对齐。

该布局水平方向是自动分块，因此不支持 match_content，布局水平宽度仅支持 match_parent 或固定宽度。

自适应仅在水平方向进行自动分块，纵向未做限制，因此如果某个子组件的高设置为 match_parent 类型，可能导致后续行无法显示，如图 4-21 所示。

1. 常用方法

AdaptiveBoxLayout 布局常用方法及功能描述如表 4-8 所示。

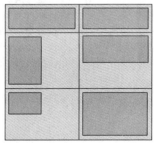

图 4-21 设置高度

表 4-8 AdaptiveBoxLayout 布局常用方法及功能描述

常用方法	功能描述
addAdaptiveRule(int minWidth,int maxWidth,int columns)	添加一个自适应盒子布局规则
removeAdaptiveRule(int minWidth,int maxWidth,int columns)	移除一个自适应盒子布局规则
clearAdaptiveRules()	移除所有自适应盒子布局规则

2. 场景示例

在 AdaptiveBoxLayout 中添加和删除自适应盒子布局规则的相关代码如下。

1) XML 布局示例代码

```xml
<?xml version = "1.0" encoding = "utf-8"?>
<DirectionalLayout
    xmlns:ohos = "http://schemas.huawei.com/res/ohos"
    ohos:height = "match_parent"
    ohos:width = "match_parent"
    ohos:orientation = "vertical">
    <AdaptiveBoxLayout
        xmlns:ohos = "http://schemas.huawei.com/res/ohos"
        ohos:height = "0vp"
        ohos:width = "match_parent"
        ohos:weight = "1"
        ohos:id = "$+id:adaptive_box_layout">
        <Text
            ohos:height = "40vp"
            ohos:width = "80vp"
            ohos:background_element = "#EC9DAA"
            ohos:margin = "10vp"
            ohos:padding = "10vp"
            ohos:text = "NO 1"
            ohos:text_size = "18fp" />
        <Text
            ohos:height = "40vp"
            ohos:width = "80vp"
            ohos:background_element = "#EC9DAA"
            ohos:margin = "10vp"
            ohos:padding = "10vp"
            ohos:text = "NO 2"
            ohos:text_size = "18fp" />
        <Text
            ohos:height = "match_content"
            ohos:width = "match_content"
            ohos:background_element = "#EC9DAA"
            ohos:margin = "10vp"
            ohos:padding = "10vp"
            ohos:multiple_lines = "true"
```

```
            ohos:text = "AdaptiveBoxLayout, where a number of boxes with the same width but
varied heights are laid out. The height of a row is determined by the highest box."
            ohos:text_size = "18fp" />
    < Text
        ohos:height = "40vp"
        ohos:width = "80vp"
        ohos:background_element = " # EC9DAA"
        ohos:margin = "10vp"
        ohos:padding = "10vp"
        ohos:text = "NO 4"
        ohos:text_size = "18fp" />
    < Text
        ohos:height = "40vp"
        ohos:width = "match_parent"
        ohos:background_element = " # EC9DAA"
        ohos:margin = "10vp"
        ohos:padding = "10vp"
        ohos:text = "Add"
        ohos:text_size = "18fp" />
    < Text
        ohos:height = "40vp"
        ohos:width = "80vp"
        ohos:background_element = " # EC9DAA"
        ohos:margin = "10vp"
        ohos:padding = "10vp"
        ohos:text = "NO 5"
        ohos:text_size = "18fp" />
    < Text
        ohos:height = "160vp"
        ohos:width = "80vp"
        ohos:background_element = " # EC9DAA"
        ohos:margin = "10vp"
        ohos:padding = "10vp"
        ohos:text = "NO 6"
        ohos:text_size = "18fp" />
</AdaptiveBoxLayout >
< Button
    ohos:id = " $ + id:add_rule_btn"
    ohos:layout_alignment = "horizontal_center"
    ohos:top_margin = "10vp"
    ohos:padding = "10vp"
    ohos:background_element = " # A9CFF0"
    ohos:height = "match_content"
    ohos:width = "match_content"
    ohos:text_size = "22fp"
    ohos:text = "adaptiveBoxLayout.addAdaptiveRule(100, 2000, 3);"/>
< Button
```

```xml
            ohos:id = " $ + id:remove_rule_btn"
            ohos:padding = "10vp"
            ohos:top_margin = "10vp"
            ohos:layout_alignment = "horizontal_center"
            ohos:bottom_margin = "10vp"
            ohos:background_element = " # D5D5D5"
            ohos:height = "match_content"
            ohos:width = "match_content"
            ohos:text_size = "22fp"
            ohos:text = "adaptiveBoxLayout.removeAdaptiveRule(100, 2000, 3);"/>
</DirectionalLayout>
```

2) Java 关键代码

```java
AdaptiveBoxLayout adaptiveBoxLayout = (AdaptiveBoxLayout)findComponentById(ResourceTable.Id
_adaptive_box_layout);
findComponentById(ResourceTable.Id_add_rule_btn).setClickedListener((component -> {
    //添加规则
    adaptiveBoxLayout.addAdaptiveRule(100, 2000, 3);
    //更新布局
    adaptiveBoxLayout.postLayout();
}));
findComponentById(ResourceTable.Id_remove_rule_btn).setClickedListener((component -> {
    //移除规则
    adaptiveBoxLayout.removeAdaptiveRule(100, 2000, 3);
    //更新布局
    adaptiveBoxLayout.postLayout();
}));
```

4.5 自定义组件与布局

HarmonyOS 提供了一套复杂且强大的 Java UI 框架，其中 Component 提供内容显示，它是界面中所有组件的基类。ComponentContainer 作为容器容纳 Component 或 ComponentContainer 对象，并对它们进行布局。

Java UI 框架也提供了一部分 Component 和 ComponentContainer 的具体子类，即常用的组件（如 Text、Button、Image 等）和常用的布局（如 DirectionalLayout、DependentLayout 等）。如果现有的组件和布局无法满足设计需求（如仿遥控器的圆盘按钮、可滑动的环形控制器等），可以通过自定义组件和自定义布局实现。

自定义组件是由开发者定义的具有一定特性的组件，通过扩展 Component 或其子类实现，可以精确控制屏幕元素的外观，也可响应用户的单击、触摸、长按等操作。

自定义布局是由开发者定义的具有特定布局规则的容器类组件，通过扩展 ComponentContainer 或其子类实现，可以将各子组件摆放到指定的位置，也可响应用户的滑动、拖曳等事件。

4.5.1 自定义组件

当 Java UI 框架提供的组件无法满足设计需求时，可以创建自定义组件，根据设计需求添加绘制任务，并定义组件的属性及事件响应，完成组件的自定义。

1. 常用接口

Component 类主要接口及功能描述如表 4-9 所示。

表 4-9 Component 类主要接口及功能描述

主 要 接 口	功 能 描 述
setEstimateSizeListener	设置测量组件的侦听器
setEstimatedSize	设置测量的宽度和高度
onEstimateSize	测量组件的大小以确定宽度和高度
EstimateSpec.getChildSizeWithMode	基于指定的大小和模式为子组件创建度量规范
EstimateSpec.getSize	从提供的度量规范中提取大小
EstimateSpec.getMode	获取该组件的显示模式
addDrawTask	添加绘制任务
onDraw	通过绘制任务更新组件时调用

2. 实现自定义组件

以自定义圆环组件为例介绍通用配置方法：在屏幕中绘制圆环，并实现单击改变圆环颜色的功能。

（1）创建自定义组件的类，并继承 Component 或其子类，添加构造方法，相关代码如下。

```
public class CustomComponent extends Component{
    public CustomComponent(Context context) {
        this(context, null);
    }
    //如需支持 XML 创建自定义组件,必须添加该构造方法
    public CustomComponent(Context context, AttrSet attrSet) {
        super(context, attrSet);
    }
}
```

（2）实现 Component.EstimateSizeListener 接口，在 onEstimateSize 方法中进行组件测量，并通过 setEstimatedSize 方法通知组件，相关代码如下。

```
public class CustomComponent extends Component implements Component.EstimateSizeListener {
    //240 为组件默认大小
    public int width = 240;
    public int height = 240;
    public CustomComponent(Context context, AttrSet attrSet) {
```

```
        ...
        //设置测量组件的侦听器
        setEstimateSizeListener(this);
    }
    ...
    @Override
    public boolean onEstimateSize(int widthEstimateConfig, int heightEstimateConfig) {
        int widthSpce = EstimateSpec.getMode(widthEstimateConfig);
        int heightSpce = EstimateSpec.getMode(heightEstimateConfig);
        int widthConfig = 0;
        switch (widthSpce) {
            case EstimateSpec.UNCONSTRAINT:
            case EstimateSpec.PRECISE:
                width = EstimateSpec.getSize(widthEstimateConfig);
                widthConfig = EstimateSpec.getSizeWithMode(width, EstimateSpec.PRECISE);
                break;
            case EstimateSpec.NOT_EXCEED:
                widthConfig = EstimateSpec.getSizeWithMode(width, EstimateSpec.PRECISE);
                break;
            default:
                break;
        }
        int heightConfig = 0;
        switch (heightSpce) {
            case EstimateSpec.UNCONSTRAINT:
            case EstimateSpec.PRECISE:
                height = EstimateSpec.getSize(heightEstimateConfig);
                heightConfig = EstimateSpec.getSizeWithMode(height, EstimateSpec.PRECISE);
                break;
            case EstimateSpec.NOT_EXCEED:
                heightConfig = EstimateSpec.getSizeWithMode(height, EstimateSpec.PRECISE);
                break;
            default:
                break;
        }
        setEstimatedSize(widthConfig, heightConfig);
        return true;
    }
}
```

自定义组件测量出的大小需通过 setEstimatedSize 通知组件,并且必须返回 true 使测量值生效。setEstimatedSize 方法的入参携带模式信息,可使用 Component.EstimateSpec.getChildSizeWithMode 方法进行拼接。测量组件的宽高需要携带模式信息,不同测量模式下的结果不同,需要根据实际需求选择适合的测量模式,如表 4-10 所示。

表 4-10 测量模式信息

模 式	作 用
UNCONSTRAINT	父组件对子组件没有约束，表示子组件可以任意大小
PRECISE	父组件已确定子组件的大小
NOT_EXCEED	已为子组件确定了最大尺寸，子组件不能超过指定尺寸

（3）自定义 XML 属性，通过构造方法中携带的参数 attrSet，可以获取在 XML 中配置的属性值，并应用在该自定义组件中，相关代码如下。

```java
public class CustomComponent extends Component implements Component.EstimateSizeListener {
    private static final String ATTR_RING_WIDTH = "ring_width";
    private static final String ATTR_RING_RADIUS = "ring_radius";
    private static final String ATTR_DEFAULT_COLOR = "default_color";
    private static final String ATTR_PRESSED_COLOR = "pressed_color";
    public float ringWidth = 20f;                          //圆环宽度
    public float ringRadius = 100f;                        //圆环半径
    public Color defaultColor = Color.YELLOW;   //默认颜色
    public Color pressedColor = Color.CYAN;     //按压态颜色
    public CustomComponent(Context context, AttrSet attrSet) {
      ...
    //初始化 XML 属性
        initAttrSet(attrSet);
    }
    private void initAttrSet(AttrSet attrSet) {
        if (attrSet == null) return;
        if (attrSet.getAttr(ATTR_DEFAULT_COLOR).isPresent()) {
            defaultColor = attrSet.getAttr(ATTR_DEFAULT_COLOR).get().getColorValue();
        }
        if (attrSet.getAttr(ATTR_RING_WIDTH).isPresent()) {
            ringWidth = attrSet.getAttr(ATTR_RING_WIDTH).get().getDimensionValue();
        }
        if (attrSet.getAttr(ATTR_RING_RADIUS).isPresent()) {
            ringRadius = attrSet.getAttr(ATTR_RING_RADIUS).get().getDimensionValue();
        }
        if (attrSet.getAttr(ATTR_PRESSED_COLOR).isPresent()) {
            pressedColor = attrSet.getAttr(ATTR_PRESSED_COLOR).get().getColorValue();
        }
    }
}
```

（4）实现 Component.DrawTask 接口，在 onDraw 方法中执行绘制任务，此方法提供画布 Canvas，可以精确控制屏幕元素的外观。在执行绘制任务之前，需要定义画笔 Paint，相关代码如下。

```java
public class CustomComponent extends Component implements Component.DrawTask, Component.
```

```java
EstimateSizeListener {
    //绘制圆环画笔
    private Paint circlePaint;
    public CustomComponen(Context context, AttrSet attrSet) {
        ...
        //初始化画笔
        initPaint();
        //添加绘制任务
        addDrawTask(this);
    }
    private void initPaint(){
        circlePaint = new Paint();
        circlePaint.setColor(defaultColor);
        circlePaint.setStrokeWidth(ringWidth);
        circlePaint.setStyle(Paint.Style.STROKE_STYLE);
    }
    @Override
    public void onDraw(Component component, Canvas canvas) {
        int x = width / 2;
        int y = height / 2;
        canvas.drawCircle(x, y, ringRadius, circlePaint);
    }
    ...
}
```

（5）实现 Component.TouchEventListener 或其他事件的接口，使组件响应用户输入，相关代码如下。

```java
public class CustomComponent extends Component implements Component.DrawTask, Component.EstimateSizeListener, Component.TouchEventListener {
    ...
    public CustomComponent(Context context, AttrSet attrSet) {
        ...
        //设置 TouchEvent 响应事件
        setTouchEventListener(this);
    }
    ...
    @Override
    public boolean onTouchEvent(Component component, TouchEvent touchEvent) {
        switch (touchEvent.getAction()) {
            case TouchEvent.PRIMARY_POINT_DOWN:
                circlePaint.setColor(pressedColor);
                invalidate();
                break;
            case TouchEvent.PRIMARY_POINT_UP:
                circlePaint.setColor(defaultColor);
                invalidate();
```

```
            break;
    }
    return true;
    }
}
```

注：需要更新 UI 显示时，可调用 invalidate()方法，示例中展示 TouchEventListener 为响应触摸事件，除此之外还可实现 ClickedListener 响应单击事件、LongClickedListener 响应长按事件等。

（6）在 XML 文件中创建并配置自定义组件。

```xml
<?xml version = "1.0" encoding = "utf-8"?>
<DirectionalLayout
    xmlns:ohos = "http://schemas.huawei.com/res/ohos"
    xmlns:custom = "http://schemas.huawei.com/res/custom"
    ohos:height = "match_parent"
    ohos:width = "match_parent"
    ohos:orientation = "vertical">
    <!-- 请根据实际包名和文件路径引入 -->
    <com.huawei.harmonyosdemo.custom.CustomComponent
        ohos:height = "300vp"
        ohos:width = "match_parent"
        ohos:background_element = "black"
        ohos:clickable = "true"
        custom:default_color = "gray"
        custom:pressed_color = "red"
        custom:ring_width = "20vp"
        custom:ring_radius = "120vp"/>
</DirectionalLayout>
```

4.5.2 自定义布局

当 Java UI 框架提供的布局无法满足需求时，根据需求创建自定义布局及规则。

1. 常用接口

下面介绍 Component 和 ComponentContainer 类相关接口。

（1）Component 类主要接口及功能描述如表 4-11 所示。

表 4-11 Component 类主要接口及功能描述

主要接口	功能描述
setEstimateSizeListener	设置测量组件的侦听器
setEstimatedSize	设置测量的宽度和高度
onEstimateSize	测量组件的大小以确定宽度和高度
EstimateSpec.getChildSizeWithMode	基于指定的大小和模式为子组件创建度量规范
EstimateSpec.getSize	从提供的度量规范中提取大小

续表

主 要 接 口	功 能 描 述
EstimateSpec.getMode	获取该组件的显示模式
Arrange	相对于容器组件设置组件的位置和大小

（2）ComponentContainer 类主要接口及功能描述如表 4-12 所示。

表 4-12　ComponentContainer 类主要接口及功能描述

主 要 接 口	功 能 描 述
setArrangeListener	设置容器组件布局子组件的侦听器
onArrange	通知容器组件在布局时设置子组件的位置和大小

2. 实现自定义布局

使用自定义布局，实现子组件自动换行功能，如图 4-22 所示。

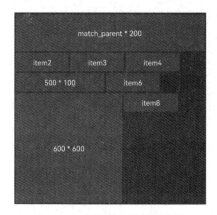

图 4-22　自定义布局使用效果

（1）创建自定义布局的类，并继承 ComponentContainer，添加构造方法如下。

```
public class CustomLayout extends ComponentContainer {
    public CustomLayout(Context context) {
        this(context, null);
    }
    //如需支持 XML 创建自定义布局,必须添加该构造方法
    public CustomLayout(Context context, AttrSet attrSet) {
        super(context, attrSet);
    }
}
```

（2）实现 ComponentContainer.EstimateSizeListener 接口，在 onEstimateSize 方法中进行测量，相关代码如下。

```
public class CustomLayout extends ComponentContainer
```

```java
        implements ComponentContainer.EstimateSizeListener {
    ...
    public CustomLayout(Context context, AttrSet attrSet) {
        ...
        setEstimateSizeListener(this);
    }
    @Override
    public boolean onEstimateSize(int widthEstimatedConfig, int heightEstimatedConfig) {
        invalidateValues();
        //通知子组件进行测量
        measureChildren(widthEstimatedConfig, heightEstimatedConfig);
        //关联子组件的索引与其布局数据
        for (int idx = 0; idx < getChildCount(); idx++) {
            Component childView = getComponentAt(idx);
            addChild(childView, idx, EstimateSpec.getSize(widthEstimatedConfig));
        }
        //测量自身
        measureSelf(widthEstimatedConfig, heightEstimatedConfig);
        return true;
    }
    private void measureChildren(int widthEstimatedConfig, int heightEstimatedConfig) {
        for (int idx = 0; idx < getChildCount(); idx++) {
            Component childView = getComponentAt(idx);
            if (childView != null) {
                LayoutConfig lc = childView.getLayoutConfig();
                int childWidthMeasureSpec;
                int childHeightMeasureSpec;
                if (lc.width == LayoutConfig.MATCH_CONTENT) {
                    childWidthMeasureSpec = EstimateSpec.getSizeWithMode(lc.width, EstimateSpec.NOT_EXCEED);
                } else if (lc.width == LayoutConfig.MATCH_PARENT) {
                    int parentWidth = EstimateSpec.getSize(widthEstimatedConfig);
                    int childWidth = parentWidth - childView.getMarginLeft() - childView.getMarginRight();
                    childWidthMeasureSpec = EstimateSpec.getSizeWithMode(childWidth, EstimateSpec.PRECISE);
                } else {
                    childWidthMeasureSpec = EstimateSpec.getSizeWithMode(lc.width, EstimateSpec.PRECISE);
                }
                if (lc.height == LayoutConfig.MATCH_CONTENT) {
                    childHeightMeasureSpec = EstimateSpec.getSizeWithMode(lc.height, EstimateSpec.NOT_EXCEED);
                } else if (lc.height == LayoutConfig.MATCH_PARENT) {
                    int parentHeight = EstimateSpec.getSize(heightEstimatedConfig);
                    int childHeight = parentHeight - childView.getMarginTop() - childView.getMarginBottom();
```

```
                            childHeightMeasureSpec = EstimateSpec.getSizeWithMode(childHeight,
    EstimateSpec.PRECISE);
                        } else {
                            childHeightMeasureSpec = EstimateSpec.getSizeWithMode(lc.height,
    EstimateSpec.PRECISE);
                        }
                        childView.estimateSize(childWidthMeasureSpec, childHeightMeasureSpec);
                }
            }
        }
        private void measureSelf(int widthEstimatedConfig, int heightEstimatedConfig) {
            int widthSpce = EstimateSpec.getMode(widthEstimatedConfig);
            int heightSpce = EstimateSpec.getMode(heightEstimatedConfig);
            int widthConfig = 0;
            switch (widthSpce) {
                case EstimateSpec.UNCONSTRAINT:
                case EstimateSpec.PRECISE:
                    int width = EstimateSpec.getSize(widthEstimatedConfig);
                    widthConfig = EstimateSpec.getSizeWithMode(width, EstimateSpec.PRECISE);
                    break;
                case EstimateSpec.NOT_EXCEED:
                    widthConfig = EstimateSpec.getSizeWithMode(maxWidth, EstimateSpec.PRECISE);
                    break;
                default:
                    break;
            }
            int heightConfig = 0;
            switch (heightSpce) {
                case EstimateSpec.UNCONSTRAINT:
                case EstimateSpec.PRECISE:
                    int height = EstimateSpec.getSize(heightEstimatedConfig);
                    heightConfig = EstimateSpec.getSizeWithMode(height, EstimateSpec.PRECISE);
                    break;
                case EstimateSpec.NOT_EXCEED:
                    heightConfig = EstimateSpec.getSizeWithMode(maxHeight, EstimateSpec.PRECISE);
                    break;
                default:
                    break;
            }
            setEstimatedSize(widthConfig, heightConfig);
        }
    }
```

注：容器类组件在自定义测量过程中不仅要测量自身，也要通知各子组件进行测量。测量出的大小需通过 setEstimatedSize 通知组件，并且必须返回 true，使测量值生效。

(3) 测量时,需要确定每个子组件大小和位置的数据,并进行保存。

```java
private int xx = 0;
    private int yy = 0;
    private int maxWidth = 0;
    private int maxHeight = 0;
    private int lastHeight = 0;
    //子组件索引与其布局数据的集合
    private final Map< Integer, Layout > axis = new HashMap<>();
    private static class Layout {
        int positionX = 0;
        int positionY = 0;
        int width = 0;
        int height = 0;
    }
    ...
    private void invalidateValues() {
        xx = 0;
        yy = 0;
        maxWidth = 0;
        maxHeight = 0;
        axis.clear();
    }
    private void addChild(Component component, int id, int layoutWidth) {
        Layout layout = new Layout();
        layout.positionX = xx + component.getMarginLeft();
        layout.positionY = yy + component.getMarginTop();
        layout.width = component.getEstimatedWidth();
        layout.height = component.getEstimatedHeight();
        if ((xx + layout.width) > layoutWidth) {
            xx = 0;
            yy += lastHeight;
            lastHeight = 0;
            layout.positionX = xx + component.getMarginLeft();
            layout.positionY = yy + component.getMarginTop();
        }
        axis.put(id, layout);
        lastHeight = Math.max(lastHeight, layout.height + component.getMarginBottom());
        xx += layout.width + component.getMarginRight();
        maxWidth = Math.max(maxWidth, layout.positionX + layout.width + component.getMarginRight());
        maxHeight = Math.max(maxHeight, layout.positionY + layout.height + component.getMarginBottom());
    }
```

(4) 实现 ComponentContainer.ArrangeListener 接口,在 onArrange 方法中排列子组件。

```java
public class CustomLayout extends ComponentContainer
```

```java
        implements ComponentContainer.EstimateSizeListener,
        ComponentContainer.ArrangeListener {
    ...
    public CustomLayout(Context context
, AttrSet attrSet
) {
        ...
        setArrangeListener(this);
    }
    @Override
    public boolean onArrange(int left, int top, int width, int height) {
        //对各子组件进行布局
        for (int idx = 0; idx < getChildCount(); idx++) {
            Component childView = getComponentAt(idx);
            Layout layout = axis.get(idx);
            if (layout != null) {
                childView.arrange(layout.positionX, layout.positionY, layout.width, layout.height);
            }
        }
        return true;
    }
}
```

(5) 在 XML 文件中创建此布局,并添加若干子组件,相关代码如下。

```xml
<?xml version = "1.0" encoding = "utf-8"?>
<DirectionalLayout
    xmlns:ohos = "http://schemas.huawei.com/res/ohos"
    ohos:height = "match_parent"
    ohos:width = "match_parent"
    ohos:orientation = "vertical">
    <!-- 请根据实际包名与文件路径引入 -->
    <com.huawei.harmonyosdemo.custom.CustomLayout
        ohos:height = "match_content"
        ohos:width = "match_parent"
        ohos:background_element = "#555555">
        <Text
            ohos:height = "200"
            ohos:width = "match_parent"
            ohos:background_element = "#727272"
            ohos:margin = "10"
            ohos:text = "match_parent * 200"
            ohos:text_alignment = "center"
            ohos:text_color = "white"
            ohos:text_size = "40"/>
        <Text
            ohos:height = "100"
```

```
        ohos:width = "300"
        ohos:background_element = "#727272"
        ohos:margin = "10"
        ohos:text = "item2"
        ohos:text_alignment = "center"
        ohos:text_color = "white"
        ohos:text_size = "40"/>
    <Text
        ohos:height = "100"
        ohos:width = "300"
        ohos:background_element = "#727272"
        ohos:margin = "10"
        ohos:text = "item3"
        ohos:text_alignment = "center"
        ohos:text_color = "white"
        ohos:text_size = "40"/>
    <Text
        ohos:height = "100"
        ohos:width = "300"
        ohos:background_element = "#727272"
        ohos:margin = "10"
        ohos:text = "item4"
        ohos:text_alignment = "center"
        ohos:text_color = "white"
        ohos:text_size = "40"/>
    <Text
        ohos:height = "100"
        ohos:width = "500"
        ohos:background_element = "#727272"
        ohos:margin = "10"
        ohos:text = "500 * 100"
        ohos:text_alignment = "center"
        ohos:text_color = "white"
        ohos:text_size = "40"/>
    <Text
        ohos:height = "100"
        ohos:width = "300"
        ohos:background_element = "#727272"
        ohos:margin = "10"
        ohos:text = "item6"
        ohos:text_alignment = "center"
        ohos:text_color = "white"
        ohos:text_size = "40"/>
    <Text
        ohos:height = "600"
        ohos:width = "600"
        ohos:background_element = "#727272"
```

```
            ohos:margin = "10"
            ohos:text = "600 * 600"
            ohos:text_alignment = "center"
            ohos:text_color = "white"
            ohos:text_size = "40"/>
        < Text
            ohos:height = "100"
            ohos:width = "300"
            ohos:background_element = " #727272"
            ohos:margin = "10"
            ohos:text = "item8"
            ohos:text_alignment = "center"
            ohos:text_color = "white"
            ohos:text_size = "40"/>
    </com.huawei.harmonyosdemo.custom.CustomLayout>
</DirectionalLayout>
```

4.6 动画开发

动画是组件的基础特性之一,精心设计的动画使 UI 变化更直观,有助于改进应用程序的外观并提升用户体验。Java UI 框架提供了帧动画、数值动画和属性动画,并提供了将多个动画同时操作的动画集合。

4.6.1 帧动画

帧动画利用视觉暂留现象,将一系列静止的图片按序播放,给用户产生动画的效果。

(1) 在 Project 窗口中,打开 entry→src→main→resources→base→media,添加一系列图片至 media 目录下。

(2) 在 graphic 目录下,新建 animation_element.xml 文件,在 XML 文件中使用 animation-list 标签配置图片资源,duration 设置显示时长,单位为 ms。oneshot 表示是否只播放一次,相关代码如下。

```
<?xml version = "1.0" encoding = "utf-8"?>
< animation - list xmlns:ohos = "http://schemas.huawei.com/res/ohos"
                  ohos:oneshot = "false">
    < item ohos:element = " $media:01" ohos:duration = "100"/>
    < item ohos:element = " $media:02" ohos:duration = "100"/>
    < item ohos:element = " $media:03" ohos:duration = "100"/>
    < item ohos:element = " $media:04" ohos:duration = "100"/>
    < item ohos:element = " $media:05" ohos:duration = "100"/>
    < item ohos:element = " $media:06" ohos:duration = "100"/>
    < item ohos:element = " $media:07" ohos:duration = "100"/>
    < item ohos:element = " $media:08" ohos:duration = "100"/>
```

```
    < item ohos:element = " $media:09" ohos:duration = "100"/>
    < item ohos:element = " $media:10" ohos:duration = "100"/>
    < item ohos:element = " $media:11" ohos:duration = "100"/>
    < item ohos:element = " $media:12" ohos:duration = "100"/>
</animation - list >
```

（3）在 MainAbilitySlice.java 中实现动画播放的相关功能。

```
import ohos.aafwk.ability.AbilitySlice;
import ohos.aafwk.content.Intent;
import ohos.agp.components.ComponentContainer;
import ohos.agp.components.DirectionalLayout;
import ohos.agp.components.Image;
import ohos.agp.components.element.FrameAnimationElement;
public class MainAbilitySlice extends AbilitySlice {
    @Override
    public void onStart(Intent intent) {
        super.onStart(intent);
        //加载动画资源,生成动画对象
        FrameAnimationElement frameAnimationElement = new FrameAnimationElement(getContext(),
ResourceTable.Graphic_animation_element);
        //创建播放动画组件
        Image image = new Image(getContext());
        image.setLayoutConfig(new ComponentContainer.LayoutConfig(500, 500));
        image.setBackground(frameAnimationElement);
        DirectionalLayout directionalLayout = new DirectionalLayout(getContext());
        directionalLayout.addComponent(image);
        super.setUIContent(directionalLayout);
        //开始播放动画
        frameAnimationElement.start();
    }
}
```

4.6.2　数值动画

AnimatorValue 数值在 0～1 变化,本身与 Component 无关。开发者可以设置 0～1 变化过程的属性,例如时长、变化曲线、重复次数等,并通过值的变化改变组件的属性,实现动画效果。

1. Java 代码方式

MainAbilitySlice.java 相关代码如下。

```
import ohos.aafwk.ability.AbilitySlice;
import ohos.aafwk.content.Intent;
import ohos.agp.animation.Animator;
import ohos.agp.animation.AnimatorValue;
import ohos.agp.components.ComponentContainer;
```

```java
import ohos.agp.components.DirectionalLayout;
import ohos.agp.components.Image;
public class MainAbilitySlice extends AbilitySlice {
    @Override
    public void onStart(Intent intent) {
        super.onStart(intent);
        //创建播放动画的组件
        Image image = new Image(getContext());
        image.setPixelMap(ResourceTable.Media_icon);
        image.setLayoutConfig(new ComponentContainer.LayoutConfig(200, 200));
        DirectionalLayout layout = new DirectionalLayout(getContext());
        layout.setLayoutConfig(new ComponentContainer.LayoutConfig(
                ComponentContainer.LayoutConfig.MATCH_PARENT,
                ComponentContainer.LayoutConfig.MATCH_PARENT
        ));
        layout.addComponent(image);
        super.setUIContent(layout);
        //创建数值动画对象
        AnimatorValue animatorValue = new AnimatorValue();
        //动画时长
        animatorValue.setDuration(3000);
        //播放前的延迟时间
        animatorValue.setDelay(1000);
        //循环次数
        animatorValue.setLoopedCount(AnimatorValue.INFINITE);
        //动画的播放类型
        animatorValue.setCurveType(Animator.CurveType.BOUNCE);
        //设置动画过程
        animatorValue.setValueUpdateListener(new AnimatorValue.ValueUpdateListener() {
            @Override
            public void onUpdate(AnimatorValue animatorValue, float value) {
                image.setContentPosition((int) (800 * value), image.getContentPositionY());
            }
        });
        //开始启动动画
        animatorValue.start();
    }
}
```

2. XML 方式

本部分介绍 animator_value.xml 和 MainAbilitySlice.java 的相关代码。

(1) 在 resources/base/animation 目录下新建 animator_value.xml 的 XML 文件。目前 XML 方式只支持 delay 和 duration 属性，animator_value.xml 相关代码如下。

```xml
<?xml version = "1.0" encoding = "UTF-8" ?>
< animator xmlns:ohos = "http://schemas.huawei.com/res/ohos"
```

```xml
        ohos:delay = "1000"
        ohos:duration = "3000"/>
```

(2) MainAbilitySlice.java 的相关代码如下。

```java
import ohos.aafwk.ability.AbilitySlice;
import ohos.aafwk.content.Intent;
import ohos.agp.animation.Animator;
import ohos.agp.animation.AnimatorProperty;
import ohos.agp.animation.AnimatorScatter;
import ohos.agp.animation.AnimatorValue;
import ohos.agp.components.Component;
import ohos.agp.components.ComponentContainer;
import ohos.agp.components.DirectionalLayout;
import ohos.agp.components.Image;
public class MainAbilitySlice extends AbilitySlice {
    @Override
    public void onStart(Intent intent) {
        super.onStart(intent);
        //创建播放动画的组件
        Image image = new Image(getContext());
        image.setPixelMap(ResourceTable.Media_icon);
        image.setLayoutConfig(new ComponentContainer.LayoutConfig(200, 200));
        DirectionalLayout layout = new DirectionalLayout(getContext());
        layout.setLayoutConfig(new ComponentContainer.LayoutConfig(
                ComponentContainer.LayoutConfig.MATCH_PARENT,
                ComponentContainer.LayoutConfig.MATCH_PARENT
        ));
        layout.addComponent(image);
        super.setUIContent(layout);
        //创建数值动画对象
        AnimatorScatter scatter = AnimatorScatter.getInstance(getContext());
        Animator animator = scatter.parse(ResourceTable.Animation_animator_value);
        if (animator instanceof AnimatorValue) {
            AnimatorValue animatorValue = (AnimatorValue) animator;
            //循环次数
            animatorValue.setLoopedCount(AnimatorValue.INFINITE);
            //动画的播放类型
            animatorValue.setCurveType(Animator.CurveType.BOUNCE);
            //设置动画过程
            animatorValue.setValueUpdateListener(new AnimatorValue.ValueUpdateListener() {
                @Override
                public void onUpdate(AnimatorValue animatorValue, float value) {
                    image.setContentPosition((int)(800 * value), image.getContentPositionY());
                }
            });
            //开始启动动画
```

```
                animatorValue.start();
            }
        }
    }
```

3. 属性动画

为 Component 的属性设置动画是常见的需求,Java UI 框架可以为 Component 设置某个属性或多个属性的动画。

1) Java 方式

MainAbilitySlice.java 的相关代码如下。

```
import ohos.aafwk.ability.AbilitySlice;
import ohos.aafwk.content.Intent;
import ohos.agp.animation.Animator;
import ohos.agp.animation.AnimatorProperty;
import ohos.agp.animation.AnimatorValue;
import ohos.agp.components.Component;
import ohos.agp.components.ComponentContainer;
import ohos.agp.components.DirectionalLayout;
import ohos.agp.components.Image;
public class MainAbilitySlice extends AbilitySlice {
    private boolean started = false;
    @Override
    public void onStart(Intent intent) {
        super.onStart(intent);
        //创建播放动画的组件
        Image image = new Image(getContext());
        image.setPixelMap(ResourceTable.Media_icon);
        image.setLayoutConfig(new ComponentContainer.LayoutConfig(200, 200));
        DirectionalLayout layout = new DirectionalLayout(getContext());
        layout.setLayoutConfig(new ComponentContainer.LayoutConfig(
                ComponentContainer.LayoutConfig.MATCH_PARENT,
                ComponentContainer.LayoutConfig.MATCH_PARENT
        ));
        layout.addComponent(image);
        super.setUIContent(layout);
        //创建属性动画对象
        AnimatorProperty animatorProperty = new AnimatorProperty();
        animatorProperty.setTarget(image);
        animatorProperty
                //x轴从100移动到800位置
                .moveFromX(100).moveToX(800)
                //透明度从0.5变化到1.0
                .alphaFrom(0.5f).alpha(1.0f)
                //旋转720°
                .rotate(720)
```

```
            //时长3s
            .setDuration(3000)
            //延迟1s
            .setDelay(1000)
            //无限循环
            .setLoopedCount(AnimatorValue.INFINITE)
            //反弹力效果
            .setCurveType(Animator.CurveType.BOUNCE);
        //单击图片开始/停止动画
        image.setClickedListener(new Component.ClickedListener() {
            @Override
            public void onClick(Component component) {
                if (!started) {
                    //开始动画
                    animatorProperty.start();
                } else {
                    //停止动画
                    animatorProperty.stop();
                }
                started = !started;
            }
        });
    }
}
```

2) XML方式

本部分介绍 animator_property.xml 和 MainAbilitySlice.java 的相关代码。

(1) 在 resources/base/animation 文件夹下声明名为 animator_property.xml 的 XML 文件。目前 XML 方式只支持 delay 和 duration 属性，animator_property.xml 相关代码如下。

```
<?xml version = "1.0" encoding = "UTF-8" ?>
<animatorProperty xmlns:ohos = "http://schemas.huawei.com/res/ohos"
                  ohos:delay = "1000"
                  ohos:duration = "3000"/>
```

(2) MainAbilitySlice.java 相关代码如下。

```
import ohos.aafwk.ability.AbilitySlice;
import ohos.aafwk.content.Intent;
import ohos.agp.animation.Animator;
import ohos.agp.animation.AnimatorProperty;
import ohos.agp.animation.AnimatorScatter;
import ohos.agp.animation.AnimatorValue;
import ohos.agp.components.Component;
import ohos.agp.components.ComponentContainer;
import ohos.agp.components.DirectionalLayout;
```

```java
import ohos.agp.components.Image;
public class MainAbilitySlice extends AbilitySlice {
    private boolean started = false;
    @Override
    public void onStart(Intent intent) {
        super.onStart(intent);
        //创建播放动画的组件
        Image image = new Image(getContext());
        image.setPixelMap(ResourceTable.Media_icon);
        image.setLayoutConfig(new ComponentContainer.LayoutConfig(200, 200));
        DirectionalLayout layout = new DirectionalLayout(getContext());
        layout.setLayoutConfig(new ComponentContainer.LayoutConfig(
                ComponentContainer.LayoutConfig.MATCH_PARENT,
                ComponentContainer.LayoutConfig.MATCH_PARENT
        ));
        layout.addComponent(image);
        super.setUIContent(layout);
        //创建属性动画对象
        AnimatorScatter scatter = AnimatorScatter.getInstance(getContext());
        Animator animator = scatter.parse(ResourceTable.Animation_animator_property);
        if (animator instanceof AnimatorProperty) {
            AnimatorProperty animatorProperty = (AnimatorProperty) animator;
            animatorProperty.setTarget(image);
            animatorProperty
                    //x 轴从 100 移动到 800 位置
                    .moveFromX(100).moveToX(800)
                    //透明度从 0.5 变化到 1.0
                    .alphaFrom(0.5f).alpha(1.0f)
                    //旋转 720°
                    .rotate(720)
                    //无限循环
                    .setLoopedCount(AnimatorValue.INFINITE)
                    //反弹力效果
                    .setCurveType(Animator.CurveType.BOUNCE);
            //单击图片开始/停止动画
            image.setClickedListener(new Component.ClickedListener() {
                @Override
                public void onClick(Component component) {
                    if (!started) {
                        //开始动画
                        animatorProperty.start();
                    } else {
                        //停止动画
                        animatorProperty.stop();
                    }
                    started = !started;
                }
```

```
            });
        }
    }
}
```

4. 动画集合

如果需要使用一个组合动画,可以把多个动画对象进行组合,并添加到 AnimatorGroup 中。AnimatorGroup 提供两种方法:runSerially() 和 runParallel(),分别表示动画按顺序开始和动画同时开始(注:动画集合暂不支持 XML 使用方式)。

(1) 多个动画同时开始。同时执行动画 1 和动画 2。动画 1:沿 x 轴从 100 移动到 800 位置。动画 2:沿 y 轴从 100 移动到 800 位置。

MainAbilitySlice.java 相关代码如下。

```
import ohos.aafwk.ability.AbilitySlice;
import ohos.aafwk.content.Intent;
import ohos.agp.animation.Animator;
import ohos.agp.animation.AnimatorGroup;
import ohos.agp.animation.AnimatorProperty;
import ohos.agp.components.Component;
import ohos.agp.components.ComponentContainer;
import ohos.agp.components.DirectionalLayout;
import ohos.agp.components.Image;
public class MainAbilitySlice extends AbilitySlice {
    private boolean started = false;
    @Override
    public void onStart(Intent intent) {
        super.onStart(intent);
        //创建播放动画的组件
        Image image = new Image(getContext());
        image.setPixelMap(ResourceTable.Media_icon);
        image.setLayoutConfig(new ComponentContainer.LayoutConfig(200, 200));
        DirectionalLayout layout = new DirectionalLayout(getContext());
        layout.setLayoutConfig(new ComponentContainer.LayoutConfig(
                ComponentContainer.LayoutConfig.MATCH_PARENT,
                ComponentContainer.LayoutConfig.MATCH_PARENT
        ));
        layout.addComponent(image);
        super.setUIContent(layout);
        //创建动画组对象
        AnimatorGroup animatorGroup = new AnimatorGroup();
        //动画1:沿 x 轴从 100 移动到 800 位置
        AnimatorProperty action1 = new AnimatorProperty();
        action1.setTarget(image);
        action1.moveFromX(0).moveToX(800);
        //动画2:沿 y 轴从 100 移动到 800 位置
```

```java
            AnimatorProperty action2 = new AnimatorProperty();
            action2.setTarget(image);
            action2.moveFromY(0).moveToY(800);
            //同时执行动画1和动画2
            animatorGroup.runParallel(action1, action2);
            //无限循环
            animatorGroup.setLoopedCount(Animator.INFINITE);
            //时长
            animatorGroup.setDuration(1500);
            //单击图片开始/停止动画
            image.setClickedListener(new Component.ClickedListener() {
                @Override
                public void onClick(Component component) {
                    if (!started) {
                        //启动动画组
                        animatorGroup.start();
                    } else {
                        //停止动画组
                        animatorGroup.stop();
                    }
                    started = !started;
                }
            });
        }
    }
```

(2) 多个动画按顺序逐个执行。先执行动画 1，然后执行动画 2。动画 1：沿 x 轴从 100 移动到 800 位置。动画 2：沿 y 轴从 100 移动到 800 位置。

MainAbilitySlice.java 相关代码如下：

```java
import ohos.aafwk.ability.AbilitySlice;
import ohos.aafwk.content.Intent;
import ohos.agp.animation.Animator;
import ohos.agp.animation.AnimatorGroup;
import ohos.agp.animation.AnimatorProperty;
import ohos.agp.components.Component;
import ohos.agp.components.ComponentContainer;
import ohos.agp.components.DirectionalLayout;
import ohos.agp.components.Image;
public class MainAbilitySlice extends AbilitySlice {
    private boolean started = false;
    @Override
    public void onStart(Intent intent) {
        super.onStart(intent);
        //创建播放动画的组件
        Image image = new Image(getContext());
        image.setPixelMap(ResourceTable.Media_icon);
```

```java
image.setLayoutConfig(new ComponentContainer.LayoutConfig(200, 200));
DirectionalLayout layout = new DirectionalLayout(getContext());
layout.setLayoutConfig(new ComponentContainer.LayoutConfig(
        ComponentContainer.LayoutConfig.MATCH_PARENT,
        ComponentContainer.LayoutConfig.MATCH_PARENT
));
layout.addComponent(image);
super.setUIContent(layout);
//创建动画组对象
AnimatorGroup animatorGroup = new AnimatorGroup();
//动画1:沿 x 轴从 100 移动到 800 位置
AnimatorProperty action1 = new AnimatorProperty();
action1.setTarget(image);
action1.moveFromX(0).moveToX(800);
//动画 2:沿 y 轴从 100 移动到 800 位置
AnimatorProperty action2 = new AnimatorProperty();
action2.setTarget(image);
action2.moveFromY(0).moveToY(800);
//先动画 1 后动画 2
animatorGroup.runSerially(action1, action2);
//无限循环
animatorGroup.setLoopedCount(Animator.INFINITE);
//时长
animatorGroup.setDuration(1500);
//单击图片开始/停止动画
image.setClickedListener(new Component.ClickedListener() {
    @Override
    public void onClick(Component component) {
        if (!started) {
            //启动动画组
            animatorGroup.start();
        } else {
            //停止动画组
            animatorGroup.stop();
        }
        started = !started;
    }
});
    }
}
```

(3) 先同时执行动画 1 和动画 2,然后同时执行动画 3 和动画 4。为了更加灵活处理多个动画的播放顺序(一些动画顺序播放,一些动画同时播放),Java UI 框架提供了更方便的动画 Builder 接口。

动画 1:沿 x 轴从 100 移动到 800 位置。动画 2:沿 y 轴从 100 移动到 800 位置。动画 3:沿 y 轴从 0.3 放大到 1.0。动画 4:沿 x 轴从 0.3 放大到 1.0。

MainAbilitySlice.java 相关代码如下。

```java
import ohos.aafwk.ability.AbilitySlice;
import ohos.aafwk.content.Intent;
import ohos.agp.animation.Animator;
import ohos.agp.animation.AnimatorGroup;
import ohos.agp.animation.AnimatorProperty;
import ohos.agp.components.Component;
import ohos.agp.components.ComponentContainer;
import ohos.agp.components.DirectionalLayout;
import ohos.agp.components.Image;
public class MainAbilitySlice extends AbilitySlice {
    private boolean started = false;
    @Override
    public void onStart(Intent intent) {
        super.onStart(intent);
        //创建播放动画的组件
        Image image = new Image(getContext());
        image.setPixelMap(ResourceTable.Media_icon);
        image.setLayoutConfig(new ComponentContainer.LayoutConfig(200, 200));
        DirectionalLayout layout = new DirectionalLayout(getContext());
        layout.setLayoutConfig(new ComponentContainer.LayoutConfig(
                ComponentContainer.LayoutConfig.MATCH_PARENT,
                ComponentContainer.LayoutConfig.MATCH_PARENT
        ));
        layout.addComponent(image);
        super.setUIContent(layout);
        //创建动画组对象
        AnimatorGroup animatorGroup = new AnimatorGroup();
        //动画1:沿 x 轴从 100 移动到 800 位置
        AnimatorProperty action1 = new AnimatorProperty();
        action1.setTarget(image);
        action1.moveFromX(0).moveToX(800);
        //动画2:沿 y 轴从 100 移动到 800 位置
        AnimatorProperty action2 = new AnimatorProperty();
        action2.setTarget(image);
        action2.moveFromY(0).moveToY(800);
        //动画3:沿 y 轴从 0.3 放大到 1.0
        AnimatorProperty action3 = new AnimatorProperty();
        action3.setTarget(image);
        action3.scaleYFrom(0.3f).scaleY(1.0f);
        //动画4:沿 x 轴从 0.3 放大到 1.0
        AnimatorProperty action4 = new AnimatorProperty();
        action4.setTarget(image);
        action4.scaleXFrom(0.3f).scaleX(1.0f);
        //首先同时执行动画 1 和动画 2,然后同时执行动画 3 和动画 4
        AnimatorGroup.Builder builder = animatorGroup.build();
```

```
builder.addAnimators(action1,action2).addAnimators(action3,action4);
//无限循环
animatorGroup.setLoopedCount(Animator.INFINITE);
//时长
animatorGroup.setDuration(1500);
//单击图片开始/停止动画
image.setClickedListener(new Component.ClickedListener() {
    @Override
    public void onClick(Component component) {
        if (!started) {
            //启动动画组
            animatorGroup.start();
        } else {
            //停止动画组
            animatorGroup.stop();
        }
        started = !started;
    }
});
}
}
```

4.7 可见即可说开发

可见即可说开发要求 Component 与热词关联,从而达到指定的效果。例如,在浏览图片时,说出图片的名字或角标序号,从而实现打开图片的效果,该功能目前仅在智慧屏产品上支持。

1. 热词注册

开发者需要进行 Component 的热词注册,提示哪些热词是 Component 需要响应的。

(1) 构建 Component.VoiceEvent 对象,需要设置热词,中英文均可。

```
Component component = new Component(getContext());
component.VoiceEvent eventKeys = new Component.VoiceEvent("ok");
```

(2) 如果一个 Component 的同一 VoiceEvent 存在多个热词匹配,可以通过 addSynonyms 方法增加 eventKeys 的热词。

```
eventKeys.addSynonyms("确定");
```

(3) 当 Component.VoiceEvent 对象操作完成后,使用 Component 的 subscribeVoiceEvents 方法发起注册。

```
component.subscribeVoiceEvents(eventKeys);
```

(4) 如果一个 Component 有多个事件需要响应,需要创建不同的事件进行注册。

2. 事件响应

开发者完成热词注册后,需要关注对应不同热词所需要处理的事件。事件响应回调的 SpeechEvent 对象仅包含一个热词。

(1) 实现 SpeechEventListener 接口的相关代码如下。

```
private Component.SpeechEventListener speechEventListener = new Component.SpeechEventListener(){
    @Override
    public boolean onSpeechEvent(Component v, SpeechEvent event) {
        if (event.getActionProperty().equals("ok")) {
            … //检测注册的热词,进行相应的处理
        }
        return false;
    };
}
```

(2) 通过 setSpeechEventListener 方法实现回调注册。

```
component.setSpeechEventListener(speechEventListener);
```

第 5 章 方舟开发框架(ArkUI)——基于 JS 扩展的类 Web 开发范式

方舟开发框架是一种跨设备的高性能 UI 开发框架,支持声明式编程和跨设备多态 UI,适用于手机、平板、智慧屏和智能穿戴应用开发。方舟开发框架包括基于 JS 扩展的类 Web 开发范式和 TS 扩展的声明式开发范式。本章对基于 JS 扩展的类 Web 开发范式进行描述。

5.1 开发概述

基于 JS 扩展的类 Web 开发基础功能如下。

1. 类 Web 范式编程

采用类 HTML 和 CSS Web 编程语言作为页面布局和页面样式的开发语言,页面业务逻辑则支持 ECMAScript 规范的 JavaScript 语言。方舟开发框架提供的类 Web 编程范式,可以让开发者避免编写 UI 状态切换的代码,视图配置信息更加直观。

2. 跨设备

开发框架架构上支持 UI 跨设备显示能力,运行时自动映射到不同设备类型,开发者无感知,降低多设备适配成本。

3. 高性能

开发框架包含许多核心的控件,例如列表、图片和各类容器组件等,针对声明式语法进行了渲染流程的优化。

使用基于 JS 扩展的类 Web 开发范式的方舟开发框架,包括应用层(Application)、前端框架层(Framework)、引擎层(Engine)和平台适配层(Porting Layer),如图 5-1 所示。

(1) 应用层。应用层表示开发的 FA 应用,这里的 FA 特指 JS FA 应用。

(2) 前端框架层。前端框架层主要完成前端页面解析,以及提供 MVVM(Model-View-ViewModel)开发模式、页面路由机制和自定义组件等功能。

(3) 引擎层。引擎层主要提供动画解析、DOM(Document Object Model)树构建、布局计算、渲染命令构建与绘制、事件管理等功能。

(4) 平台适配层。平台适配层主要完成对平台层进行抽象,提供抽象接口,可以对接到

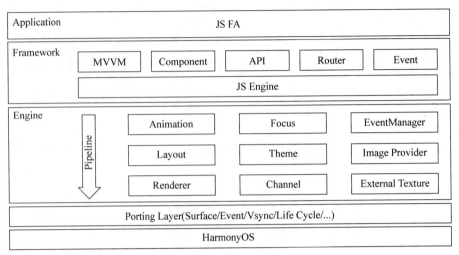

图 5-1 方舟开发框架

系统平台。例如,事件对接、渲染管线对接和系统生命周期对接等。

5.2　JS FA 初步应用

基于 JS 扩展的类 Web 开发范式支持纯 JavaScript、JavaScript 和 Java 混合语言开发。JS FA 是基于 JavaScript、JavaScript 与 Java 混合开发的 FA。

5.2.1　JS FA 概述

JS FA 在 HarmonyOS 上运行时,需要基类 AceAbility、加载 JS FA 主体的方法、JS FA 开发目录,具体说明如下。

1. AceAbility

AceAbility 类是 JS FA 在 HarmonyOS 上运行环境的基类,继承自 Ability。开发者的应用运行入口类应该从该类派生,相关代码如下。

```
public class MainAbility extends AceAbility {
    @Override
    public void onStart(Intent intent) {
        super.onStart(intent);
    }
    @Override
    public void onStop() {
        super.onStop();
    }
}
```

2. 加载 JS FA

JS FA 生命周期事件分为应用生命周期和页面生命周期,应用通过 AceAbility 类中 setInstanceName()接口设置该 Ability 的实例资源,并通过 AceAbility 窗口进行显示及全局应用生命周期管理。

setInstanceName(String name)的参数 name 是指实例名称,实例名称与 config.json 文件中 module.js.name 的值对应。若开发者未修改实例名称,而使用了默认值 default,则无须调用此接口。若已经修改,则需在应用 Ability 实例的 onStart()中调用此接口,并将参数 name 设置为修改后的实例名称。多实例应用的 module.js 字段中有多个实例项,使用时选择相应的实例名称。

setInstanceName()接口使用方法:在 MainAbility 的 onStart()中的 super.onStart()前调用此接口。以 JSComponentName 作为实例名称,需在 super.onStart(Intent)前调用此接口,相关代码如下。

```
public class MainAbility extends AceAbility {
    @Override
    public void onStart(Intent intent) {
        setInstanceName("JSComponentName");
//config.json 配置文件中 module.js.name 的标签值
        super.onStart(intent);
    }
}
```

3. JS FA 开发目录

新建工程的 JS 目录如图 5-2 所示。

在工程目录中,i18n 下存放多语言的 json 文件;pages 文件夹下存放多个页面,每个页面由 HML、CSS 和 JS 文件组成。HML 是一套类 HTML 的标记语言,通过组件、事件构建出页面的内容。页面具备数据绑定、事件绑定、列表渲染、条件渲染等高级能力。

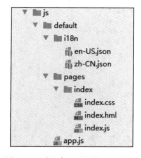

图 5-2 新建工程的 JS 目录

(1) main→js→default→i18n→en-US.json:此文件定义了在英文模式下页面显示的变量内容。

```
{
  "strings": {
    "hello": "Hello",
    "world": "World"
  }
}
```

同理,zh-CN.json 中定义了中文模式下的页面内容。

(2) main→js→default→pages→index→index.hml:此文件定义了 index 页面的布局、

用到的组件,以及这些组件的层级关系。例如,index.hml 文件中包含了一个 text 组件,内容为 Hello World 文本。

```
< div class = "container">
  < text class = "title">
    {{ $t('strings.hello') }} {{title}}
  </text>
</div>
```

(3) main→js→default→pages→index→index.css:此文件定义了 index 页面的样式(index.css 文件定义了 container 和 title)。

```
.container {
  flex-direction: column;
  justify-content: center;
  align-items: center;
}
.title {
  font-size: 100px;
}
```

(4) main→js→default→pages→index→index.js:此文件定义了 index 页面的业务逻辑(数据绑定、事件处理等)。示例:变量 title 赋值为字符串 World。

```
export default {
  data: {
    title: '',
  },
  onInit() {
    this.title = this.$t('strings.world');
  },
}
```

5.2.2　JS FA 开发应用

本节主要介绍 JS FA 开发应用。该应用通过 media query 同时适配手机和 TV,单击或者将焦点移动到食物的缩略图选择不同的食物图片,也可以添加到购物车,如图 5-3 和图 5-4 所示。

1. 构建页面布局

在 index.hml 文件中构建页面布局,进行代码开发之前,首先要对页面布局进行分析,将页面分解为不同区,用容器组件承载。根据 JS FA 应用效果图,此页面共分成三部分:标题区、展示区和详情区。展示区和详情区在手机和 TV 上分别是按列排列和按行排列。

标题区较为简单,由两个按列排列的 text 组件构成。展示区包含 4 个 image 组件的 swiper,详情区由 image 和 text 组件构成。下面以手机效果图为例,展示区和详情区布局如图 5-5 所示。

图 5-3　手机应用效果

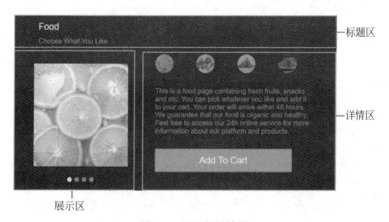

图 5-4　TV 应用效果

根据布局结构的分析，实现页面基础布局的代码如下（其中 4 个 image 组件通过 for 指令循环创建）。

```
<!-- index.hml -->
<div class = "container">
  <!-- title area -->
  <div class = "title">
    <text class = "name"> Food </text>
```

图 5-5　展示区和详情区布局

```
    < text class = "sub-title"> Choose What You Like </text >
  </div >
  < div class = "display-style">
    <!-- display area -->
    < swiper id = "swiperImage" class = "swiper-style">
      < image src = "{{$item}}" class = "image-mode" focusable = "true" for = "{{imageList}}">
</image >
    </swiper >
    <!-- product details area -->
    < div class = "container">
      < div class = "selection-bar-container">
        < div class = "selection-bar">
          < image src = "{{$item}}" class = "option-mode" onfocus = "swipeToIndex({{$idx}})"
onclick = "swipeToIndex({{$idx}})" for = "{{imageList}}"></image >
        </div >
      </div >
      < div class = "description-first-paragraph">
        < text class = "description">{{descriptionFirstParagraph}}</text >
      </div >
      < div class = "cart">
        < text class = "{{cartStyle}}" onclick = "addCart" onfocus = "getFocus" onblur =
"lostFocus" focusable = "true">{{cartText}}</text >
      </div >
    </div >
  </div >
</div >
```

swiper组件里展示的图片需要自行添加图片资源,放置到js→default→common目录下,common目录需自行创建。

2. 构建页面样式

index.css文件通过media query管控手机和TV不同页面样式。此外,该页面样式还采用了CSS伪类的写法,当单击或者焦点移动到image组件上时,image组件由半透明变成不透明,以此实现选中的效果,相关代码请扫描二维码获取。

3. 构建页面逻辑

在index.js文件中构建页面逻辑,主要实现两个功能:单击或者焦点移动到不同的缩略图时,swiper滑动到相应的图片;焦点移动到购物车时,Add To Cart背景颜色从浅蓝色变成深蓝色,单击后文字变为Cart+1,背景颜色由深蓝色变成黄色。添加购物车不可重复操作,逻辑页面相关代码如下。

```js
//index.js
export default {
  data: {
    cartText: 'Add To Cart',
    cartStyle: 'cart-text',
    isCartEmpty: true,
    descriptionFirstParagraph: 'This is a food page containing fresh fruits, snacks and etc. You can pick whatever you like and add it to your cart. Your order will arrive within 48 hours. We guarantee that our food is organic and healthy. Feel free to access our 24h online service for more information about our platform and products.',
    imageList: ['/common/food_000.JPG', '/common/food_001.JPG', '/common/food_002.JPG', '/common/food_003.JPG'],
  },
  swipeToIndex(index) {
    this.$element('swiperImage').swipeTo({index: index});
  },
  addCart() {
    if (this.isCartEmpty) {
      this.cartText = 'Cart + 1';
      this.cartStyle = 'add-cart-text';
      this.isCartEmpty = false;
    }
  },
  getFocus() {
    if (this.isCartEmpty) {
      this.cartStyle = 'cart-text-focus';
    }
  },
  lostFocus() {
    if (this.isCartEmpty) {
      this.cartStyle = 'cart-text';
    }
  },
}
```

4. 配置设备类型

在config.json的deviceType字段中添加手机和TV的设备类型。

```
{
    ...
    "module": {
        ...
        "deviceType": [
            "phone",
            "tv"
        ],
        ...
    }
}
```

5.3 构建用户界面

本节主要介绍组件、构建布局、添加交互、动画、事件、页面路由和焦点逻辑。

5.3.1 组件

组件(Component)是构建页面的核心,每个组件通过对数据和方法的简单封装,实现独立的可视、可交互功能单元。组件之间相互独立,随取随用,既可以在需求相同的地方重复使用,也可以通过组件间合理的搭配定义满足业务需求的新组件,减少开发量,实现自定义开发的组件。组件根据功能,可以分为以下几类,如表5-1所示。

表5-1 组件类型及组件名称

组件类型	组件名称
容器组件	badge、dialog、div、form、list、list-item、list-item-group、panel、popup、refresh、stack、stepper、stepper-item、swiper、tabs、tab-bar、tab-content
基础组件	button、chart、divider、image、image-animator、input、label、marquee、menu、option、picker、picker-view、piece、progress、qrcode、rating、richtext、search、select、slider、span、switch、text、textarea、toolbar、toolbar-item、toggle、web
媒体组件	camera、video
画布组件	canvas
栅格组件	grid-container、grid-row、grid-col
svg组件	svg、rect、circle、ellipse、path、line、polyline、polygon、text、tspan、textPath、animate、animateMotion、animateTransform

5.3.2 构建布局

本节主要对布局说明、添加标题行和文本区域、添加图片区域、添加留言区域、添加容器进行描述。

1. 布局说明

手机和智慧屏的基准宽度为720px(px为逻辑像素,非物理像素),实际显示效果会根据屏幕宽度进行缩放,其换算关系如下。

组件的width设为100px时,在宽度为720物理像素的屏幕上,实际显示为100物理像素;在宽度为1440物理像素的屏幕上,实际显示为200物理像素。智能穿戴的基准宽度为454px,换算逻辑同理。

一个页面的基本元素包含标题区域、文本区域和图片区域等,每个基本元素内还包含多个子元素,根据需求可以添加按钮、开关、进度条等组件。在构建页面布局时,需要对每个基本元素思考以下几个问题:元素的尺寸和排列位置、是否有重叠的元素、是否需要设置对齐、内间距或者边界、是否包含子元素及其排列位置、是否需要容器组件及其类型。

将页面中的元素分解之后再对每个基本元素按顺序实现,可以减少多层嵌套造成的视觉混乱和逻辑混乱,提高代码的可读性,方便对页面做后续调整,如图5-6和图5-7所示。

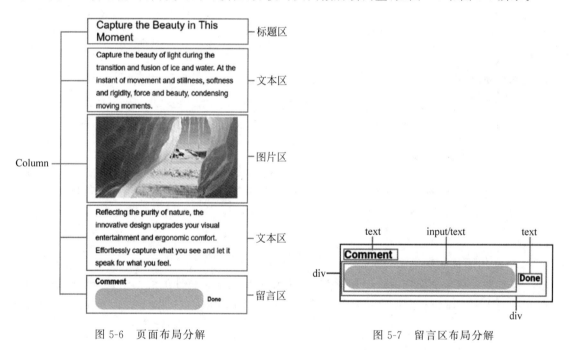

图5-6 页面布局分解　　　　　图5-7 留言区布局分解

2. 添加标题行和文本区域

实现标题和文本区域常用的是基础组件text。text组件用于展示文本,可以设置不同的属性和样式,文本内容需要写在标签内容区,插入标题和文本区域的相关代码如下:

```
<!-- xxx.hml -->
<div class = "container">
  <text class = "title-text">{{headTitle}}</text>
```

```html
      <text class = "paragraph-text">{{paragraphFirst}}</text>
      <text class = "paragraph-text">{{paragraphSecond}}</text>
</div>
```
```css
/* xxx.css */
.container {
  flex-direction: column;
  margin-top: 20px;
  margin-left: 30px;
}
.title-text {
  color: #1a1a1a;
  font-size: 50px;
  margin-top: 40px;
  margin-bottom: 20px;
}
.paragraph-text {
  color: #000000;
  font-size: 35px;
  line-height: 60px;
}
```
```js
//xxx.js
export default {
  data: {
    headTitle: 'Capture the Beauty in This Moment',
    paragraphFirst: 'Capture the beauty of light during the transition and fusion of ice and water. At the instant of movement and stillness, softness and rigidity, force and beauty, condensing moving moments.',
    paragraphSecond: 'Reflecting the purity of nature, the innovative design upgrades your visual entertainment and ergonomic comfort. Effortlessly capture what you see and let it speak for what you feel.',
  },
}
```

3. 添加图片区域

添加图片区域通常用 image 组件实现,使用方法与 text 组件类似。图片资源建议放在 js→default→common 目录下,common 目录需自行创建,相关代码如下。

```html
<!-- xxx.hml -->
<image class = "img" src = "{{middleImage}}"></image>
```
```css
/* xxx.css */
.img {
  margin-top: 30px;
  margin-bottom: 30px;
  height: 385px;
}
```
```js
//xxx.js
export default {
```

```
  data: {
    middleImage: '/common/ice.png',
  },
}
```

4. 添加留言区域

用户输入留言后单击完成,留言区域即显示留言内容。用户单击右侧的"删除"按钮可删除当前留言内容并重新输入,留言区域由 div、text、input 关联 click 事件实现。开发者可以使用 input 组件实现输入留言,通过 text 组件实现留言完成,使用 commentText 的状态标记此时显示的组件(通过 if 属性控制)。在包含文本"完成"和"删除"的 text 组件中关联 click 事件,更新 commentText 状态和 inputValue 的内容,相关代码如下。

```html
<!-- xxx.hml -->
<div class = "container">
  <text class = "comment-title"> Comment </text>
  <div if = "{{!commentText}}">
    <input class = "comment" value = "{{inputValue}}" onchange = "updateValue()"></input>
    <text class = "comment-key" onclick = "update" focusable = "true"> Done </text>
  </div>
  <div if = "{{commentText}}">
    <text class = "comment-text" focusable = "true">{{inputValue}}</text>
    <text class = "comment-key" onclick = "update" focusable = "true"> Delete </text>
  </div>
</div>
```

```css
/* xxx.css */
.container {
  margin-top: 24px;
  background-color: #ffffff;
}
.comment-title {
  font-size: 40px;
  color: #1a1a1a;
  font-weight: bold;
  margin-top: 40px;
  margin-bottom: 10px;
}
.comment {
  width: 550px;
  height: 100px;
  background-color: lightgrey;
}
.comment-key {
  width: 150px;
  height: 100px;
  margin-left: 20px;
  font-size: 32px;
```

```css
    color: #1a1a1a;
    font-weight: bold;
}
.comment-key:focus {
    color: #007dff;
}
.comment-text {
    width: 550px;
    height: 100px;
    text-align: left;
    line-height: 35px;
    font-size: 30px;
    color: #000000;
    border-bottom-color: #bcbcbc;
    border-bottom-width: 0.5px;
}
```

```js
//xxx.js
export default {
    data: {
        inputValue: '',
        commentText: false,
    },
    update() {
        this.commentText = !this.commentText;
    },
    updateValue(e) {
        this.inputValue = e.text;
    },
}
```

5. 添加容器

将页面的基本元素组装在一起,需要使用容器组件。在页面布局中常用div、list和tabs 3种容器组件。在页面结构相对简单时,可以直接用div作为容器,因为div作为单纯的布局容器,可以支持多种子组件,使用起来更为方便。

1) list 组件

当页面结构较为复杂时,如果使用div循环渲染,容易出现卡顿,因此推荐使用list组件代替div组件实现长列表布局,从而实现更加流畅的列表滚动体验。list仅支持list-item作为子组件,相关代码如下。

```html
<!-- xxx.hml -->
<list class="list">
    <list-item type="listItem" for="{{textList}}">
        <text class="desc-text">{{$item.value}}</text>
    </list-item>
</list>
```

```css
/* xxx.css */
.desc-text {
  width: 683.3px;
  font-size: 35.4px;
}
```
```js
//xxx.js
export default {
  data: {
    textList: [{value: 'JS FA'}],
  },
}
```

为避免示例代码过长，以上示例的 list 中只包含一个 list-item 和一个 text 组件。在实际应用中可以在 list 中加入多个 list-item，同时 list-item 下可以包含多个其他子组件。

2）tabs 组件

当页面经常需要动态加载时，推荐使用 tabs 组件。tabs 组件支持 change 事件，在页签切换后触发。tabs 组件仅支持一个 tab-bar 和一个 tab-content，相关代码如下。

```html
<!-- xxx.hml -->
<tabs>
  <tab-bar>
    <text>Home</text>
    <text>Index</text>
    <text>Detail</text>
  </tab-bar>
  <tab-content>
    <image src="{{homeImage}}"></image>
    <image src="{{indexImage}}"></image>
    <image src="{{detailImage}}"></image>
  </tab-content>
</tabs>
```
```js
//xxx.js
export default {
  data: {
    homeImage: '/common/home.png',
    indexImage: '/common/index.png',
    detailImage: '/common/detail.png',
  },
}
```

tab-content 组件用来展示页签的内容区，支持 scrollable 属性，高度默认充满 tabs 剩余空间。

5.3.3 添加交互

添加交互可以通过在组件上实现关联事件。本节介绍如何用 div、text 和 image 组件关联 click 事件，构建一个点赞按钮。点赞按钮通过一个 div 组件实现关联 click 事件。div 组件包含一个 image 和一个 text 组件。

image 组件用于显示未点赞和点赞的效果。click 事件的函数会交替更新点赞和未点赞图片的路径。text 组件用于显示点赞数，点赞数会在 click 事件的函数中同步更新。

click 事件作为一个函数定义在 JS 文件中，可以更改 isPressed 的状态，从而更新显示 image 组件。如果 isPressed 为真，则点赞数加 1。该函数在 HML 文件中对应的 div 组件上生效，点赞按钮各子组件的样式设置在 CSS 文件当中，相关代码如下：

```
<!-- xxx.hml -->
<!-- 点赞按钮 -->
<div>
  <div class="like" onclick="likeClick">
    <image class="like-img" src="{{likeImage}}" focusable="true"></image>
    <text class="like-num" focusable="true">{{total}}</text>
  </div>
</div>
/* xxx.css */
.like {
  width: 104px;
  height: 54px;
  border: 2px solid #bcbcbc;
  justify-content: space-between;
  align-items: center;
  margin-left: 72px;
  border-radius: 8px;
}
.like-img {
  width: 33px;
  height: 33px;
  margin-left: 14px;
}
.like-num {
  color: #bcbcbc;
  font-size: 20px;
  margin-right: 17px;
}
//xxx.js
export default {
  data: {
    likeImage: '/common/unLike.png',
    isPressed: false,
    total: 20,
  },
  likeClick() {
    var temp;
```

```
    if (!this.isPressed) {
      temp = this.total + 1;
      this.likeImage = '/common/like.png';
    } else {
      temp = this.total - 1;
      this.likeImage = '/common/unLike.png';
    }
    this.total = temp;
    this.isPressed = !this.isPressed;
  },
}
```

除此之外,还有很多表单组件(开关、标签、滑动选择器等)可在页面布局时灵活使用,提高交互性。

5.3.4 动画

动画分为静态动画和连续动画。

1. 静态动画

静态动画的核心是 transform 样式,可以实现以下 3 种变换类型,一次样式设置只能实现一种类型变换。

translate:沿水平或垂直方向将指定组件移动所需距离。scale:横向或纵向将指定组件缩小或放大到所需比例。rotate:将指定组件沿横轴、纵轴或中心点旋转指定的角度,相关代码如下。

```
<!-- xxx.hml -->
<div class = "container">
  <text class = "translate">hello</text>
  <text class = "rotate">hello</text>
  <text class = "scale">hello</text>
</div>
/* xxx.css */
.container {
  flex-direction: column;
  align-items: center;
}
.translate {
  height: 150px;
  width: 300px;
  font-size: 50px;
  background-color: #008000;
  transform: translate(200px);
}
.rotate {
  height: 150px;
```

```
    width: 300px;
    font-size: 50px;
    background-color: #008000;
    transform-origin: 200px 100px;
    transform: rotateX(45deg);
}
.scale {
    height: 150px;
    width: 300px;
    font-size: 50px;
    background-color: #008000;
    transform: scaleX(1.5);
}
```

2. 连续动画

静态动画只有开始状态和结束状态,没有中间状态。如果设置中间的过渡状态和转换效果,需要使用连续动画实现。

连续动画的核心是 animation 样式,它定义了动画的开始状态、结束状态及时间和速度的变化曲线,通过 animation 样式实现的效果如下。

animation-name:设置动画执行后应用到组件上的背景颜色、透明度、宽高和变换类型。

animation-delay 和 animation-duration:分别设置动画执行后元素延迟和持续的时间。

animation-timing-function:描述动画执行的速度曲线,使动画更加平滑。

animation-iteration-count:定义动画播放的次数。

animation-fill-mode:指定动画执行结束后是否恢复初始状态。

animation 样式需要在 CSS 文件中先定义 keyframe,然后设置动画的过渡效果,并通过一个样式类型在 HML 文件中调用,相关代码如下。

```
<!-- xxx.hml -->
<div class="item-container">
  <text class="header">animation-name</text>
  <div class="item {{colorParam}}">
    <text class="txt">color</text>
  </div>
  <div class="item {{opacityParam}}">
    <text class="txt">opacity</text>
  </div>
  <input class="button" type="button" name="" value="show" onclick="showAnimation"/>
</div>
/* xxx.css */
.item-container {
    margin-right: 60px;
    margin-left: 60px;
    flex-direction: column;
```

```css
}
.header {
  margin-bottom: 20px;
}
.item {
  background-color: #f76160;
}
.txt {
  text-align: center;
  width: 200px;
  height: 100px;
}
.button {
  width: 200px;
  font-size: 30px;
  background-color: #09ba07;
}
.color {
  animation-name: Color;
  animation-duration: 8000ms;
}
.opacity {
  animation-name: Opacity;
  animation-duration: 8000ms;
}
@keyframes Color {
  from {
    background-color: #f76160;
  }
  to {
    background-color: #09ba07;
  }
}
@keyframes Opacity {
  from {
    opacity: 0.9;
  }
  to {
    opacity: 0.1;
  }
}
```

```js
//xxx.js
export default {
  data: {
    colorParam: '',
    opacityParam: '',
  },
```

```
showAnimation: function () {
    this.colorParam = '';
    this.opacityParam = '';
    this.colorParam = 'color';
    this.opacityParam = 'opacity';
},
}
```

5.3.5 事件

事件主要为手势事件和按键事件。手势事件主要用于智能穿戴等具有触摸屏的设备，按键事件主要用于智慧屏设备。

1. 手势事件

手势表示由单个或多个事件识别的语义动作（触摸、单击和长按）。一个完整的手势可能由多个事件组成，对应手势的生命周期，支持的事件如下。

1) 触摸

touchstart：手指触摸动作开始。

touchmove：手指触摸后移动。

touchcancel：手指触摸动作被打断（来电提醒、弹窗）。

touchend：手指触摸动作结束。

2) 单击

click：用户快速轻敲屏幕。

3) 长按

longpress：用户在相同位置长时间保持与屏幕接触，相关代码如下。

```
<!-- xxx.hml -->
<div class = "container">
  <div class = "text-container" onclick = "click">
    <text class = "text-style">{{onClick}}</text>
  </div>
  <div class = "text-container" ontouchstart = "touchStart">
    <text class = "text-style">{{touchstart}}</text>
  </div>
  <div class = "text-container" ontouchmove = "touchMove">
    <text class = "text-style">{{touchmove}}</text>
  </div>
  <div class = "text-container" ontouchend = "touchEnd">
    <text class = "text-style">{{touchend}}</text>
  </div>
  <div class = "text-container" ontouchcancel = "touchCancel">
    <text class = "text-style">{{touchcancel}}</text>
  </div>
  <div class = "text-container" onlongpress = "longPress">
```

```html
      <text class = "text-style">{{onLongPress}}</text>
    </div>
</div>
```
```css
/* xxx.css */
.container {
  flex-direction: column;
  justify-content: center;
  align-items: center;
}
.text-container {
  margin-top: 10px;
  flex-direction: column;
  width: 750px;
  height: 50px;
  background-color: #09ba07;
}
.text-style {
  width: 100%;
  line-height: 50px;
  text-align: center;
  font-size: 24px;
  color: #ffffff;
}
```
```js
//xxx.js
export default {
  data: {
    touchstart: 'touchstart',
    touchmove: 'touchmove',
    touchend: 'touchend',
    touchcancel: 'touchcancel',
    onClick: 'onclick',
    onLongPress: 'onlongpress',
  },
  touchCancel: function (event) {
    this.touchcancel = 'canceled';
  },
  touchEnd: function(event) {
    this.touchend = 'ended';
  },
  touchMove: function(event) {
    this.touchmove = 'moved';
  },
  touchStart: function(event) {
    this.touchstart = 'touched';
  },
  longPress: function() {
    this.onLongPress = 'longpressed';
```

```
      },
      click: function() {
        this.onClick = 'clicked';
      },
    }
```

2. 按键事件

按键事件是智慧屏上特有的手势事件,当用户操作遥控器按键时触发。用户单击一个遥控器按键,通常会触发两次 key 事件:先触发 action 为 0,再触发 action 为 1,即先触发按下事件,再触发抬起事件。action 为 2 的场景比较少见,一般为用户按下按键且不松开,此时 repeatCount 将返回次数。每个物理按键对应各自的按键值(keycode)实现不同的功能,相关代码如下。

```
<!-- xxx.hml -->
<div class = "card-box">
  <div class = "content-box">
    <text class = "content-text" onkey = "keyUp" onfocus = "focusUp" onblur = "blurUp">{{up}}
</text>
  </div>
  <div class = "content-box">
    <text class = "content-text" onkey = "keyDown" onfocus = "focusDown" onblur = "blurDown">
{{down}}</text>
  </div>
</div>
/* xxx.css */
.card-box {
  flex-direction: column;
  justify-content: center;
}
.content-box {
  align-items: center;
  height: 200px;
  flex-direction: column;
  margin-left: 200px;
  margin-right: 200px;
}
.content-text {
  font-size: 40px;
  text-align: center;
}
//xxx.js
export default {
  data: {
    up: 'up',
    down: 'down',
  },
```

```
    focusUp: function() {
      this.up = 'up focused';
    },
    blurUp: function() {
      this.up = 'up';
    },
    keyUp: function() {
      this.up = 'up keyed';
    },
    focusDown: function() {
      this.down = 'down focused';
    },
    blurDown: function() {
      this.down = 'down';
    },
    keyDown: function() {
      this.down = 'down keyed';
    },
}
```

按键事件通过获焦事件向下分发,因此示例中使用了 focus 事件和 blur 事件明确当前焦点的位置。单击上下键选中 up 或 down 按键,即相应的 focused 状态,失去焦点的按键恢复正常的 up 或 down 按键文本,按确认键后该按键变为 keyed 状态。

5.3.6 页面路由

很多应用由多个页面组成,例如,用户可以从音乐列表页面单击歌曲,跳转到该歌曲的播放界面。此过程通过页面路由进行串联,按需实现跳转。

页面路由 router 根据页面的 URI 找到目标页面,从而实现跳转。以最基础的两个页面之间的跳转为例,具体实现步骤如下:在 Project 窗口,打开 entry→src→main→js→default,右击 pages 文件夹,选择 New→JS Page,创建详情页;调用 router.push()路由到详情页,调用 router.back()回到首页。

1. 构建页面布局

index 和 detail 页面均包含一个 text 组件和 button 组件。text 组件指明当前页面,button 组件实现两个页面之间的相互跳转,HML 文件相关代码如下。

```
<!-- index.hml -->
<div class="container">
  <text class="title">This is the index page.</text>
  <button type="capsule" value="Go to the second page" class="button" onclick="launch">
  </button>
</div>
<!-- detail.hml -->
<div class="container">
```

```html
    <text class = "title">This is the detail page.</text>
    <button type = "capsule" value = "Go back" class = "button" onclick = "launch"></button>
</div>
```

2. 构建页面样式

构建 index 和 detail 页面样式，text 组件和 button 组件居中显示，两个组件间距为 50px。CSS 相关代码如下（两个页面样式代码一致）。

```
/* index.css */
/* detail.css */
.container {
  flex-direction: column;
  justify-content: center;
  align-items: center;
}
.title {
  font-size: 50px;
  margin-bottom: 50px;
}
```

3. 实现跳转

为了使 button 组件的 launch 方法生效，需要在页面的 JS 文件中实现跳转逻辑。调用 router.push()接口，将 URI 指定的页面添加到路由栈中，即跳转到 URI 指定的页面。在调用 router 方法之前，需要导入 router 模块，相关代码如下。

```
//index.js
import router from '@system.router';
export default {
  launch() {
    router.push({
      uri: 'pages/detail/detail',
    });
  },
}
//detail.js
import router from '@system.router';
export default {
  launch() {
    router.back();
  },
}
```

运行效果如图 5-8 所示。

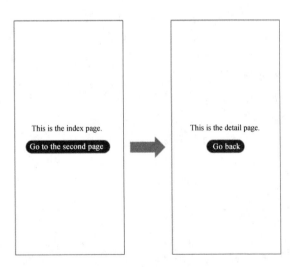

图 5-8　页面路由效果

5.3.7 焦点逻辑

焦点移动是智慧屏的主要交互方式,本节介绍焦点逻辑的相关规则。

1. 容器组件焦点分发逻辑

容器组件在第一次获焦时焦点一般落在第一个可获焦的子组件上,再次获焦时焦点落在上一次失去焦点时获焦的子组件上。容器组件一般都有特定的焦点分发逻辑,常用容器组件的焦点分发逻辑说明如下。

(1) div 组件通过按键移动获焦时,焦点会移动到在移动方向上与当前获焦组件布局中心距离最近的可获焦叶子节点上。如图 5-9 所示,焦点在上方横向 div 的第二个子组件上,当单击 Down 按键时,焦点要移动到下方的横向 div 中,这时下方横向 div 中的子组件会与当前焦点所在的子组件进行布局中心距离的计算,其中距离最近的子组件获焦。

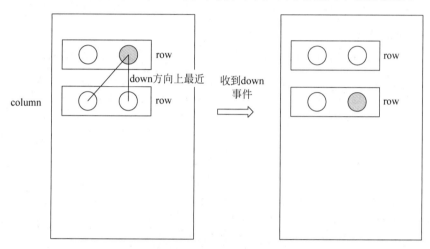

图 5-9 div 焦点移动时距离计算示例

(2) list 组件包含 list-item 与 list-item-group,list 组件每次获焦时会使第一个可获焦的 item 获焦。list-item-group 为特殊的 list-item,且两者都与 div 的焦点逻辑相同。

(3) stack 组件只能由自顶而下的第一个可获焦的子组件获焦。

(4) swiper 的每个页面和 refresh 页面焦点逻辑都与 div 的相同。

(5) tabs 组件包含 tab-bar 与 tab-content,tab-bar 中的子组件默认均能获焦,与是否有可获焦的叶子节点无关。tab-bar 与 tab-content 的每个页面都与 div 的焦点逻辑相同。

(6) dialog 的 button 可获焦,若有多个 button,默认初始焦点落在第二个 button 上。

(7) popup 无法获焦。

2. focusable 属性使用

通用属性 focusable 主要用于控制组件能否获焦,本身不支持焦点的组件在设置此属性后可以拥有获取焦点的能力。例如,text 组件本身不能获焦,焦点无法移动到它上面,设置 text 的 focusable 属性为 true 后,text 组件便可以获焦。如果在未使用 focusable 属性的情况下,使用了 focus、blur 或 key 事件,会默认添加 focusable 属性为 true。

容器组件能否可获焦依赖于是否拥有可获焦的子组件。如果容器组件内没有可以获焦的子组件,即使设置了 focusable 为 true,依然不能获焦。若容器组件 focusable 属性设置为 false,则它本身和所包含的所有组件都不可获焦。

5.4 常见组件开发

常见组件包括 Text、Input、Button、List、Picker、Dialog、Form、Stepper、Tabs、Image。下面分别介绍功能及其相关代码。

5.4.1 Text

Text 是文本组件,用于呈现一段文本信息。

1. 创建 Text 组件

在 pages/index 目录下的 HML 文件中创建一个 Text 组件,相关代码如下。

```
<!-- xxx.hml -->
<div class="container" style="text-align: center;justify-content: center; align-items: center;">
  <text>
    Hello World
  </text>
</div>
/* xxx.css */
.container {
  flex-direction: column;
  align-items: center;
  justify-content: center;
  background-color: #F1F3F5;
}
```

2. 设置 Text 组件样式和属性

设置 Text 组件样式和属性步骤如下。

1) 添加文本样式

设置 color、font-size 和 allow-scale 属性分别为文本添加颜色、大小和缩放,相关代码如下。

```
<!-- xxx.hml -->
<div class="container" style="background-color: #F1F3F5; justify-content: center; align-items: center;">
  <text style="color: blueviolet; font-size: 40px; allow-scale:true">
    This is a passage
  </text>
</div>
```

2）添加画线

设置 text-decoration 属性为文本添加画线，相关代码如下。

```html
<!-- xxx.hml -->
<div class="container" style="background-color:#F1F3F5;">
  <text style="text-decoration:underline">
    This is a passage
  </text>
  <text style="text-decoration:line-through">
    This is a passage
  </text>
</div>
```

```css
/* xxx.css */
.container {
  flex-direction: column;
  align-items: center;
  justify-content: center;
}
text{
  font-size: 50px;
}
```

3）隐藏文本内容

当文本内容过多而显示不全时，添加 text-overflow 属性将隐藏内容以省略号的形式展现，相关代码如下。

```html
<!-- xxx.hml -->
<div class="container">
  <text class="text">
    This is a passage
  </text>
</div>
```

```css
/* xxx.css */
.container {
  flex-direction: column;
  align-items: center;
  background-color: #F1F3F5;
  justify-content: center;
}
.text{
  width: 200px;
  max-lines: 1;
  text-overflow:ellipsis;
}
```

text-overflow 样式需要与 max-lines 样式配套使用，max-lines 属性设置文本最多可以展示的行数,在设置最大行数的情况下方可生效。

4) 设置文本折行

设置 word-break 属性对文本内容做断行处理,相关代码如下。

```html
<!-- xxx.hml -->
<div class = "container">
  <div class = "content">
    <text class = "text1">
      Welcome to the world
    </text>
    <text class = "text2">
      Welcome to the world
    </text>
  </div>
</div>
```

```css
/* xxx.css */
.container {
  background-color: #F1F3F5;
  flex-direction: column;
  align-items: center;
  justify-content: center;
}
.content{
  width: 50%;
  flex-direction: column;
  align-items: center;
  justify-content: center;
}
.text1{
  height: 200px;
  border:1px solid #1a1919;
  margin-bottom: 50px;
  text-align: center;
  word-break: break-word;
  font-size: 40px;
}
.text2{
  height: 200px;
  border:1px solid #0931e8;
  text-align: center;
  word-break: break-all;
  font-size: 40px;
}
```

5) Text 组件支持 Span 子组件

相关代码如下。

```html
<!-- xxx.hml -->
<div class="container" style="justify-content: center; align-items: center; flex-direction: column;background-color: #F1F3F5;">
  <text style="font-size: 45px;">
    This is a passage
  </text>
  <text style="font-size: 45px;">
    <span style="color: aqua;">This </span>
    <span style="color: #F1F3F5;">
      1
    </span>
    <span style="color: blue;"> is a </span>
    <span style="color: #F1F3F5;">
      1
    </span>
    <span style="color: red;"> passage </span>
  </text>
</div>
```

当使用 Span 子组件组成文本段落时，如果 Span 属性样式异常（font-weight 设置为 1000），将导致文本段落显示异常。在使用 Span 子组件时，Text 组件内不能存在文本内容；如果存在，只会显示子组件 Span 中的内容。

3. 场景示例

Text 组件通过数据绑定展示文本内容，Span 组件通过设置 show 属性实现文本内容的隐藏和显示，相关代码如下。

```html
<!-- xxx.hml -->
<div class="container">
  <div style="align-items: center;justify-content: center;">
    <text class="title">
      {{ content }}
    </text>
    <switch checked="true" onchange="test"></switch>
  </div>
  <text class="span-container" style="color: #ff00ff;">
    <span show="{{isShow}}">  {{ content }}  </span>
    <span style="color: white;">
      1
    </span>
    <span style="color: #f76160"> Hide clip </span>
  </text>
</div>
```
```css
/* xxx.css */
.container {
  align-items: center;
  flex-direction: column;
```

```css
    justify-content: center;
    background-color: #F1F3F5;
}
.title {
    font-size: 26px;
    text-align:center;
    width: 200px;
    height: 200px;
}
```
```js
//xxx.js
export default {
    data: {
        isShow:true,
        content: 'Hello World'
    },
    onInit(){    },
    test(e) {
        this.isShow = e.checked
    }
}
```

5.4.2　Input

Input 是交互式组件,用于接收用户数据,其类型可设置为日期、多选框和按钮等。

1. 创建 Input 组件

在 pages/index 目录下的 HML 文件中创建 Input 组件,相关代码如下。

```html
<!-- xxx.hml -->
<div class="container">
  <input type="text">
    Please enter the content
  </input>
</div>
```
```css
/* xxx.css */
.container {
    flex-direction: column;
    justify-content: center;
    align-items: center;
    background-color: #F1F3F5;
}
```

2. 设置 Input 类型

通过设置 type 属性定义 Input 类型,将 Input 设置为 button、date 等,相关代码如下。

```html
<!-- xxx.hml -->
<div class="container">
```

```html
    <div class = "div-button">
      <dialog class = "dialogClass" id = "dialogId">
        <div class = "content">
          <text> this is a dialog </text>
        </div>
      </dialog>
      <input class = "button" type = "button" value = "click" onclick = "btnclick"></input>
    </div>
    <div class = "content">
      <input onchange = "checkboxOnChange" checked = "true" type = "checkbox"></input>
    </div>
    <div class = "content">
      <input type = "date" class = "flex" placeholder = "Enter data"></input>
    </div>
</div>
```
```css
/* xxx.css */
.container {
  align-items: center;
  flex-direction: column;
  justify-content: center;
  background-color: #F1F3F5;
}
.div-button {
  flex-direction: column;
  align-items: center;
}
.dialogClass{
  width:80%;
  height: 200px;
}
.button {
  margin-top: 30px;
  width: 50%;
}
.content{
  width: 90%;
  height: 150px;
  align-items: center;
  justify-content: center;
}
.flex {
  width: 80%;
  margin-bottom:40px;
}
```
```js
//xxx.js
export default {
  btnclick(){
```

```
        this.$element('dialogId').show()
    },
}
```

智能穿戴将 Input 类型设置为 button、radio 和 checkbox。当 Input 类型为 checkbox 和 radio 时,当前组件选中的属性为 checked 才生效,默认值为 false。

3. 事件绑定

向 Input 组件添加 search 和 translate 事件,相关代码如下。

```
<!-- xxx.hml -->
<div class="content">
  <text style="margin-left: -7px;">
    <span>Enter text and then touch and hold what you've entered</span>
  </text>
  <input class="input" type="text" onsearch="search" placeholder="search"></input>
  <input class="input" type="text" ontranslate="translate" placeholder="translate"></input>
</div>
/* xxx.css */
.content {
  width: 100%;
  flex-direction: column;
  align-items: center;
  justify-content: center;
  background-color: #F1F3F5;
}
.input {
  margin-top: 50px;
  width: 60%;
  placeholder-color: gray;
}
text{
  width:100%;
  font-size:25px;
  text-align:center;
}
// xxx.js
import prompt from '@system.prompt'
export default {
  search(e){
    prompt.showToast({
      message:  e.value,
      duration: 3000,
    });
  },
  translate(e){
```

```
      prompt.showToast({
        message: e.value,
        duration: 3000,
      });
    }
  }
```

4. 设置输入提示

通过对 Input 组件添加 showError 方法提示输入的错误原因,相关代码如下。

```
<!-- xxx.hml -->
<div class = "content">
  <input id = "input" class = "input" type = "text" maxlength = "20" placeholder = "Please input text" onchange = "change">
  </input>
  <input class = "button" type = "button" value = "Submit" onclick = "buttonClick"></input>
</div>
/* xxx.css */
.content {
  width: 100%;
  flex-direction: column;
  align-items: center;
  justify-content: center;
  background-color: #F1F3F5;
}
.input {
  width: 80%;
  placeholder-color: gray;
}
.button {
  width: 30%;
  margin-top: 50px;
}
//xxx.js
import prompt from '@system.prompt'
export default {
  data:{
    value:'',
  },
  change(e){
    this.value = e.value;
    prompt.showToast({
      message: "value: " + this.value,
      duration: 3000,
    });
  },
  buttonClick(e){
```

```
      if(this.value.length > 6){
        this.$element("input").showError({
          error: 'Up to 6 characters are allowed.'
        });
      }else if(this.value.length == 0){
        this.$element("input").showError({
          error:this.value + 'This field cannot be left empty.'
        });
      }else{
        prompt.showToast({
          message: "success "
        });
      }
    },
}
```

该方法在 Input 类型为 text、email、date、time、number 和 password 时生效。

5. 场景示例

根据场景选择不同类型的 Input 输入框,完成信息录入,相关代码如下。

```
<!-- xxx.hml -->
<div class = "container">
  <div class = "label-item">
    <label>memorandum</label>
  </div>
  <div class = "label-item">
    <label class = "lab" target = "input1">content:</label>
    <input class = "flex" id = "input1" placeholder = "Enter content" />
  </div>
  <div class = "label-item">
    <label class = "lab" target = "input3">date:</label>
    <input class = "flex" id = "input3" type = "date" placeholder = "Enter data" />
  </div>
  <div class = "label-item">
    <label class = "lab" target = "input4">time:</label>
    <input class = "flex" id = "input4" type = "time" placeholder = "Enter time" />
  </div>
  <div class = "label-item">
    <label class = "lab" target = "checkbox1">Complete:</label>
    <input class = "flex" type = "checkbox" id = "checkbox1" style = "width: 100px; height: 100px;" />
  </div>
  <div class = "label-item">
    <input class = "flex" type = "button" id = "button" value = "save" onclick = "btnclick"/>
  </div>
</div>
/* xxx.css */
```

```css
.container {
  flex-direction: column;
  background-color: #F1F3F5;
}
.label-item {
  align-items: center;
  border-bottom-width: 1px;border-color: #dddddd;
}
.lab {
  width: 400px;}
label {
  padding: 30px;
  font-size: 30px;
  width: 320px;
  font-family: serif;
  color: #9370d8;
  font-weight: bold;
}
.flex {
  flex: 1;
}
.textareaPadding {
  padding-left: 100px;
}
```

```js
//xxx.js
import prompt from '@system.prompt';
export default {
  data: {
  },
  onInit() {
  },
  btnclick(e) {
    prompt.showToast({
      message:'Saved successfully!'
    })
  }
}
```

5.4.3 Button

Button 是按钮组件,其类型包括胶囊按钮、圆形按钮、文本按钮、弧形按钮、下载按钮。

1. 创建 Button 组件

在 pages/index 目录下的 hml 文件中创建 Button 组件,相关代码如下。

```html
<!-- xxx.hml -->
<div class="container">
```

```html
    <button type = "capsule" value = "Capsule button"></button>
</div>
```
```css
/* xxx.css */
.container {
  flex-direction: column;
  justify-content: center;
  align-items: center;
  background-color: #F1F3F5;
}
```

2. 设置 Button 类型

通过设置 Button 的 type 属性选择按钮类型,例如定义 Button 为圆形按钮、文本按钮等,相关代码如下。

```html
<!-- xxx.hml -->
<div class = "container">
    <button class = "circle" type = "circle" >+</button>
    <button class = "text" type = "text"> button </button>
</div>
```
```css
/* xxx.css */
.container {
  background-color: #F1F3F5;
  flex-direction: column;
  align-items: center;
  justify-content: center;
}
.circle {
  font-size: 120px;
  background-color: blue;
  radius: 72px;
}
.text {
  margin-top: 30px;
  text-color: white;
  font-size: 30px;
  font-style: normal;
  background-color: blue;
  width: 50%;
  height: 100px;
}
```

胶囊按钮(type=capsule)不支持 border 相关样式。圆形按钮(type=circle)不支持文本相关样式。文本按钮(type=text)自适应文本大小,不支持尺寸样式设置(radius、width 和 height),背景透明不支持 background-color 样式。Button 组件使用的 icon 图标如果来自云端路径,需要添加网络访问权限 ohos.permission.INTERNET,在 resources 文件夹下的 config.json 文件中进行权限配置,相关代码如下。

第5章　方舟开发框架(ArkUI)——基于JS扩展的类Web开发范式

```
<!-- config.json -->
"module": {
  "reqPermissions": [{
    "name": "ohos.permission.INTERNET"
  }],
}
```

3. 显示下载进度

为 Button 组件添加 progress 方法，实时显示下载的进度，相关代码如下。

```
<!-- xxx.hml -->
<div class = "container">
  <button class = "button download" type = "download" id = "download-btn" onclick = "setProgress">
{{downloadText}}</button>
</div>
/* xxx.css */
.container {
  background-color: #F1F3F5;
  flex-direction: column;
  align-items: center;
  justify-content: center;
}
.download {
  width: 280px;
  text-color: white;
  background-color: #007dff;
}
//xxx.js
import prompt from '@system.prompt';
export default {
  data: {
    percent: 0,
    downloadText: "Download",
    isPaused: true,
    intervalId : null,
  },
  star(){
    this.intervalId = setInterval(() = >{
      if(this.percent < 100){
        this.percent += 1;
        this.downloadText = this.percent + " % ";
      } else{
        prompt.showToast({
          message: "Download succeeded."
        })
        this.paused()
        this.downloadText = "Download";
```

```
          this.percent = 0;
          this.isPaused = true;
        }
      },100)
    },
    paused(){
      clearInterval(this.intervalId);
      this.intervalId = null;
    },
    setProgress(e) {
      if(this.isPaused){
        prompt.showToast({
          message: "Download started"
        })
        this.star();
        this.isPaused = false;
      }else{
        prompt.showToast({
          message: "Paused."
        })
        this.paused();
        this.isPaused = true;
      }
    }
}
```

setProgress 方法只支持 Button 的类型为 download。

4. 场景示例

在本场景中,可根据输入的文本内容进行 Button 类型切换,相关代码如下。

```
<!-- xxx.hml -->
<div class = "container">
  <div class = "input - item">
    <input class = "input - text" id = "change" type = "{{mytype}}"  placeholder = "{{myholder}}"
      style = "background - color:{{mystyle1}};
      placeholder - color:{{mystyle2}};flex - grow:{{myflex}};"name = "{{myname}}" value =
"{{myvalue}}"></input>
  </div>
  <div class = "input - item">
    <div class = "doc - row">
      <input type = "button" class = "select - button color - 3" value = "text" onclick =
"changetype3"></input>
      <input type = "button" class = "select - button color - 3" value = "data" onclick =
"changetype4"></input>
    </div>
  </div>
</div>
```

```css
/* xxx.css */
.container {
  flex-direction: column;
  align-items: center;
  background-color: #F1F3F5;
}
.input-item {
  margin-bottom: 80px;
  flex-direction: column;
}
.doc-row {
  justify-content: center;
  margin-left: 30px;
  margin-right: 30px;
  justify-content: space-around;
}
.input-text {
  height: 80px;
  line-height: 80px;
  padding-left: 30px;
  padding-right: 30px;
  margin-left: 30px;
  margin-right: 30px;
  margin-top:100px;
  border: 3px solid;
  border-color: #999999;
  font-size: 30px;
  background-color: #ffffff;
  font-weight: 400;
}
.select-button {
  width: 35%;
  text-align: center;
  height: 70px;
  padding-top: 10px;
  padding-bottom: 10px;
  margin-top: 30px;
  font-size: 30px;
  color: #ffffff;
}
.color-3 {
  background-color: #0598db;;
}
```
```js
//xxx.js
export default {
  data: {
    myflex: '',
```

```
      myholder: 'Enter text.',
      myname: '',
      mystyle1: "#ffffff",
      mystyle2: "#ff0000",
      mytype: 'text',
      myvalue: '',
    },
    onInit() {
    },
    changetype3() {
      this.myflex = '';
      this.myholder = 'Enter text.';
      this.myname = '';
      this.mystyle1 = "#ffffff";
      this.mystyle2 = "#FF0000";
      this.mytype = 'text';
      this.myvalue = '';
    },
    changetype4() {
      this.myflex = '';
      this.myholder = 'Enter a date.';
      this.myname = '';
      this.mystyle1 = "#ffffff";
      this.mystyle2 = "#FF0000";
      this.mytype = 'date';
      this.myvalue = '';
    },
}
```

其他组件介绍,请扫描二维码获取。

5.5 动效开发

本节主要介绍 CSS 和 JS 动画。其中,CSS 动画包括属性样式、transform 样式、background-position 样式动画。JS 动画包括组件和插值器动画。

5.5.1 CSS 动画开发

本部分包括属性样式动画、transform 样式动画和 background-position 样式动画。

1. 属性样式动画

在关键帧(Keyframes)中动态设置父组件的 width 和 height 值实现变大缩小,子组件设置 scale 属性实现父子组件同时缩放,再设置 opacity 实现父子组件的显示与隐藏,相关代码如下:

```
<!-- xxx.hml -->
```

```html
<div class = "container">
  <div class = "fade">
    <text> fading away </text>
  </div>
  <div class = "bigger">
    <text> getting bigger </text>
  </div>
</div>
```

```css
/* xxx.css */
.container {
  background-color: #F1F3F5;
  display: flex;
  justify-content: center;
  align-items: center;
  flex-direction: column;
}
.fade{
  width: 30%;
  height: 200px;
  left: 35%;
  top: 25%;
  position: absolute;
  animation: 2s change infinite friction;
}
.bigger{
  width: 20%;
  height: 100px;
  background-color: blue;
  animation: 2s change1 infinite linear-out-slow-in;
}
text{
  width: 100%;
  height: 100%;
  text-align: center;
  color: white;
  font-size: 35px;
  animation: 2s change2 infinite linear-out-slow-in;
}
/* 颜色变化 */
@keyframes change{
  from {
    background-color: #f76160;
    opacity: 1;
  }
  to {
    background-color: #09ba07;
    opacity: 0;
```

```
      }
    }
    /*父组件大小变化*/
    @keyframes change1{
      0%{
        width: 20%;
        height: 100px;
      }
      100%{
        width: 80%;
        height: 200px;
      }
    }
    /*子组件文字缩放*/
    @keyframes change2{
      0%{
        transform: scale(0);
      }
      100%{
        transform: scale(1.5);
      }
    }
```

animation 取值不区分先后，duration（动画执行时间）/delay（动画延迟执行时间）按照出现的先后顺序解析。必须设置 animation-duration 样式，否则时长为 0 则不会有动画效果。当设置 animation-fill-mode 属性为 forwards 时，组件直接展示最后一帧的样式。

2. transform 样式动画

设置 transform 属性对组件进行旋转、缩放、移动和倾斜。

1）设置静态动画

创建一个正方形并旋转 90°变成菱形，用下方的长方形将菱形下半部分遮盖形成屋顶。首先，设置长方形 translate 属性值为(150px，－150px)，确定坐标位置形成门；其次，使用 position 属性将横纵线跟随父组件（正方形）移动到指定坐标位置；再次，设置 scale 属性使父子组件一起变大形成窗户大小；最后，使用 skewX 属性使组件倾斜后设置坐标 translate (200px，－830px)得到烟囱，相关代码如下。

```
<!-- xxx.hml -->
<div class = "container">
  <div class = "top"></div>
  <div class = "content"></div>
  <div class = "door"></div>
  <!-- 窗户 -->
  <div class = "window">
    <div class = "horizontal"></div>
    <div class = "vertical"></div>
```

```html
    </div>
    <div class = "chimney"></div>
</div>
```
```css
/* xxx.css */
.container {
  background-color: #F1F3F5;
  align-items: center;
  flex-direction: column;
}
.top{
  z-index: -1;
  position: absolute;
  width: 428px;
  height: 428px;
  background-color: #860303;
  transform: rotate(45deg);
  margin-top: 230px;
  margin-left: 266px;
}
.content{
  margin-top: 500px;
  width: 600px;
  height: 400px;
  background-color: white;
  border:  1px solid black;
}
.door{
  width: 100px;
  height: 150px;
  background-color: #1033d9;
  transform: translate(150px,-150px);
}
.window{
  z-index: 1;
  position: relative;
  width: 100px;
  height: 100px;
  background-color: white;
  border: 1px solid black;
  transform: translate(-150px,-400px) scale(1.5);
}
/* 窗户的横轴 */
.horizontal{
  position: absolute;
  top: 50%;
  width: 100px;
  height: 5px;
```

```css
      background-color: black;
    }
    /* 窗户的纵轴 */
    .vertical{
      position: absolute;
      left: 50%;
      width: 5px;
      height: 100px;
      background-color: black;
    }
    .chimney{
      z-index: -2;
      width: 40px;
      height: 100px;
      border-radius: 15px;
      background-color: #9a7404;
      transform: translate(200px, -830px) skewX(-5deg);
    }
```

2) 设置平移动画

小球下降动画,改变小球的 Y 轴坐标实现小球下落,在下一段时间内减小 Y 轴坐标实现小球回弹,让每次回弹的高度逐次减小直至回弹高度为 0,即模拟出小球下降的动画,相关代码如下。

```html
<!-- xxx.hml -->
<div class="container">
  <div class="circle"></div>
  <div class="flower"></div>
</div>
```

```css
/* xxx.css */
.container {
  background-color: #F1F3F5;
  display: flex;
  justify-content: center;
}
.circle{
  width: 100px;
  height: 100px;
  border-radius: 50px;
  background-color: red;
  /* forwards 停在动画的最后一帧 */
  animation: down 3s fast-out-linear-in forwards;
}
.flower{
  position: fixed;
  width: 80%;
```

```css
    margin-left: 10%;
    height: 5px;
    background-color: black;
    top: 1000px;
}
@keyframes down {
  0% {
    transform: translate(0px,0px);
  }
  /*下落*/
  15% {
    transform: translate(10px,900px);
  }
  /*开始回弹*/
  25% {
    transform: translate(20px,500px);
  }
  /*下落*/
  35% {
    transform: translate(30px,900px);
  }
  /*回弹*/
  45% {
    transform: translate(40px,700px);
  }
  55% {
    transform: translate(50px,900px);
  }
  65% {
    transform: translate(60px,800px);
  }
  80% {
    transform: translate(70px,900px);
  }
  90% {
    transform: translate(80px,850px);
  }
  /*停止*/
  100% {
    transform: translate(90px,900px);
  }
}
```

3) 设置旋转动画

设置不同的原点位置(transform-origin)改变元素围绕的旋转中心。rotate3d 属性前三个参数值分别为 X 轴、Y 轴、Z 轴的旋转向量,第四个值为旋转角度,旋转角度可为负值,负值代表旋转方向为逆时针,相关代码请扫描二维码获取。

transform-origin 变换对象的原点位置,如果仅设置一个值,另一个值为 50%,若设置两个值,则第一个值表示 X 轴的位置,第二个值表示 Y 轴的位置。

4) 设置缩放动画

设置 scale 样式属性实现涟漪动画。首先,使用定位确定元素的位置,确定坐标后创建多个组件实现重合效果;其次,设置 opacity 属性改变组件不透明度实现组件隐藏与显示,同时设置 scale 值,使组件可以一边放大一边隐藏;最后,设置两个组件不同的动画执行时间,实现扩散效果。设置 sacle3d 中 X 轴、Y 轴、Z 轴的缩放参数实现动画,相关代码如下。

```html
<!-- xxx.hml -->
<div class="container">
  <div class="circle">
    <text>ripple</text>
  </div>
  <div class="ripple"></div>
  <div class="ripple ripple2"></div>
  <!-- 3d -->
  <div class="content">
    <text>spring</text>
  </div>
</div>
```

```css
/* xxx.css */
.container {
  flex-direction: column;
  background-color: #F1F3F5;
  width: 100%;
  position: relative;
}
.circle{
  margin-top: 400px;
  margin-left: 40%;
  width: 100px;
  height: 100px;
  border-radius: 50px;
  background:linear-gradient(#dcaec1, #d3a8e3);
  z-index: 1;
  position: absolute;
}
.ripple{
  margin-top: 400px;
  margin-left: 40%;
  position: absolute;
  z-index: 0;
  width: 100px;
  height: 100px;
```

```css
  border-radius: 50px;
  background:linear-gradient(#dcaec1,#d3a8e3);
  animation: ripple 5s infinite;
}
/* 设置不同的动画时间 */
.ripple2{
  animation-duration: 2.5s;
}
@keyframes ripple{
  0% {
    transform: scale(1);
    opacity: 0.5;
  }
  50% {
    transform: scale(3);
    opacity: 0;
  }
  100% {
    transform: scale(1);
    opacity: 0.5;
  }
}
text{
  color: white;
  text-align: center;
  height: 100%;
  width: 100%;
}
.content {
  margin-top: 700px;
  margin-left: 33%;
  width: 200px;
  height: 100px;
  animation:rubberBand 1s infinite;
  /* 设置渐变色 */
  background:linear-gradient(#e276aa,#ec0d66);
  position: absolute;
}
@keyframes rubberBand {
  0% {
    transform: scale3d(1, 1, 1);
  }
  30% {
    transform: scale3d(1.25, 0.75, 1.1);
  }
  40% {
    transform: scale3d(0.75, 1.25, 1.2);
```

```
    }
    50% {
      transform: scale3d(1.15, 0.85, 1.3);
    }
    65% {
      transform: scale3d(.95, 1.05, 1.2);
    }
    75% {
      transform: scale3d(1.05, .95, 1.1);
    }
    100% {
      transform: scale3d(1, 1, 1);
    }
}
```

设置 transform 属性值后,子元素会随父元素一起改变,若只改变父元素其他属性值(如 height、width),则子元素不会改变。

5) 设置 matrix 属性

matrix 是一个入参为 6 个值的矩阵,6 个值分别代表 scaleX、skewY、skewX、scaleY、translateX 和 translateY。下面示例中设置了 matrix 属性为 matrix(1,0,0,1,0,200),使组件移动和倾斜,相关代码如下。

```
<!-- xxx.hml -->
<div class = "container">
  <div class = "rect"></div>
</div>
/* xxx.css */
.container{
  background-color:#F1F3F5;
  display: flex;
  justify-content: center;
}
.rect{
  width: 100px;
  height: 100px;
  background-color: red;
  animation: down 3s infinite forwards;
}
@keyframes down{
  0% {
    transform: matrix(1,0,0,1,0,0);
  }
  10% {
    transform: matrix(1,0,0,1,0,200);
  }
  60% {
```

```
      transform: matrix(2,1.5,1.5,2,0,700);
  }
  100% {
      transform: matrix(1,0,0,1,0,0);
  }
}
```

6) 整合 transform 属性

transform 可以设置多个值并且多个值可同时设置,下面示例中展示同时设置缩放(scale)、平移(translate)和旋转(rotate)属性时的动画效果。

```
<!-- xxx.hml -->
<div class = "container">
  <div class = "rect1"></div>
  <div class = "rect2"></div>
  <div class = "rect3"></div>
  <div class = "rect4"></div>
  <div class = "rect5"></div>
</div>
/* xxx.css */
.container{
  flex-direction:column;
  background-color:#F1F3F5;
  padding:50px;
}
.rect1{
  width: 100px;
  height: 100px;
  background:linear-gradient(#e77070,#ee0202);
  animation: change1 3s infinite forwards;
}
.rect2{
  margin-top: 50px;
  width: 100px;
  height: 100px;
  background:linear-gradient(#95a6e8,#2739de);
  animation: change2 3s infinite forwards;
}
.rect3{
  margin-top: 50px;
  width: 100px;
  height: 100px;
  background:linear-gradient(#142ee2,#8cb1e5);
  animation: change3 3s infinite;
}
.rect4{
  align-self: center;
```

```css
    margin-left: 50px;
    margin-top: 200px;
    width: 100px;
    height: 100px;
    background:linear-gradient(#e2a8df,#9c67d4,#8245d9,#e251c3);
    animation: change4 3s infinite;
}
.rect5{
    margin-top: 300px;
    width: 100px;
    height: 100px;
    background:linear-gradient(#e7ded7,#486ccd,#94b4d2);
    animation: change5 3s infinite;
}
/* change1 change2 对比 */
@keyframes change1{
    0%{
        transform: translate(0,0);
        transform: rotate(0deg)
    }
    100%{
        transform: translate(0,500px);
        transform: rotate(360deg)
    }
}
/* change2 change3 对比属性顺序不同的动画效果 */
@keyframes change2{
    0%{
        transform:translate(0,0) rotate(0deg) ;
    }
    100%{
        transform: translate(300px,0) rotate(360deg);
    }
}
@keyframes change3{
    0%{
        transform:rotate(0deg) translate(0,0);
    }
    100%{
        transform:rotate(360deg)  translate(300px,0);
    }
}
/* 属性值不对应的情况 */
@keyframes change4{
    0%{
        transform: scale(0.5);
    }
```

```
        100% {
            transform:scale(2) rotate(45deg);
        }
    }
    /* 多属性的写法 */
    @keyframes change5{
        0% {
            transform:scale(0) translate(0,0) rotate(0);
        }
        100% {
            transform: scale(1.5) rotate(360deg) translate(200px,0);
        }
    }
```

当设置多个 transform 时,后续的 transform 值会将前面的覆盖。若想同时使用多个动画样式可用复合写法,例如 transform：scale(1) rotate(0) translate(0,0)。transform 进行复合写法时,变化样式内多个样式值顺序的不同会呈现不同的动画效果。transform 属性设置的样式值要一一对应,若前后不一致,则该动画不生效。若设置多个样式值,则只会呈现出已对应值的动画效果。

3. background-position 样式动画

通过改变 background-position 属性(第一个值为 X 轴的位置,第二个值为 Y 轴的位置)移动背景图片位置,若背景图位置超出组件,则超出部分的背景图不显示,相关代码如下。

```
<!-- xxx.hml -->
<div class = "container">
    <div class = "content"></div>
    <div class = "content1"></div>
</div>
/* xxx.css */
.container {
    background-color: #F1F3F5;
    display: flex;
    flex-direction: column;
    justify-content: center;
    align-items: center;
    width: 100%;
}
.content{
    width: 400px;
    height: 400px;
    background-image: url('common/images/bg-tv.jpg');
    background-size: 100%;
    background-repeat: no-repeat;
    animation: change 3s infinite;
    border: 1px solid black;
```

```css
    }
    .content1{
      margin-top:50px;
      width: 400px;
      height: 400px;
      background-image: url('common/images/bg-tv.jpg');
      background-size: 50%;
      background-repeat: no-repeat;
      animation: change1 5s infinite;
      border: 1px solid black;
    }
    /*背景图片移动出组件*/
    @keyframes change{
      0%{
        background-position:0px top;
      }
      25%{
        background-position:400px top;
      }
      50%{
        background-position:0px top;
      }
      75%{
        background-position:0px bottom;
      }
      100%{
        background-position:0px top;
      }
    }
    /*背景图片在组件内移动*/
    @keyframes change1{
      0%{
        background-position:left top;
      }
      25%{
        background-position:50% 50%;
      }
      50%{
        background-position:right bottom;
      }
      100%{
        background-position:left top;;
      }
    }
```

background-position 仅支持背景图片的移动,不支持背景颜色(background-color)。

5.5.2 JS动画

本部分介绍组件动画、插值器动画和动画帧。

1. 组件动画

在组件上创建和运行动画的快捷方式步骤如下。

1) 获取动画对象

通过调用animate方法获得animation对象,animation对象支持动画属性、动画方法和动画事件,相关代码如下。

```
<!-- xxx.hml -->
<div class = "container">
  <div id = "content" class = "box" onclick = "Show"></div>
</div>
/* xxx.css */
.container {
  flex-direction: column;
  justify-content: center;
  align-items: center;
  width: 100%;
}
.box{
  width: 200px;
  height: 200px;
  background-color: #ff0000;
  margin-top: 30px;
}
/* xxx.js */
export default {
  data: {
    animation: '',
  },
  onInit() {
  },
  onShow() {
    var options = {
      duration: 1500,
    };
    var frames = [
      {
        width:200,height:200,
      },
      {
        width:300,height:300,
      }
    ];
```

```
      this.animation = this.$element('content').animate(frames, options);
      //获取动画对象
    },
    Show() {
      this.animation.play();
    }
  }
```

使用 animate 方法时必须传入 Keyframes 和 Options 参数。多次调用 animate 方法时，采用 replace 策略，即最后一次调用时传入的参数生效。

2) 设置动画参数

在获取动画对象后，通过设置参数 Keyframes 设置动画在组件上的样式，相关代码如下。

```
<!-- xxx.hml -->
<div class = "container">
  <div id = "content" class = "box" onclick = "Show"></div>
</div>
/* xxx.css */
.container {
  flex-direction: column;
  justify-content: center;
  align-items: center;
  width: 100%;
}
.box{
  width: 200px;
  height: 200px;
  background-color: #ff0000;
  margin-top: 30px;
}
/* xxx.js */
export default {
  data: {
    animation: '',
    keyframes:{},
    options:{}
  },
  onInit() {
    this.options = {
      duration: 4000,
    };
    this.keyframes = [
      {
        transform: {
          translate: '-120px -0px',
```

```
          scale: 1,
          rotate: 0
        },
        transformOrigin: '100px 100px',
        offset: 0.0,
        width: 200,
        height: 200
      },
      {
        transform: {
          translate: '120px 0px',
          scale: 1.5,
          rotate: 90
        },
        transformOrigin: '100px 100px',
        offset: 1.0,
        width: 300,
        height: 300
      }
    ];
  },
  Show() {
    this.animation = this.$element('content').animate(this.keyframes, this.options);
    this.animation.play();
  }
}
```

translate、scale 和 totate 的先后顺序会影响动画效果。transformOrigin 只对 scale 和 totate 起作用。在获取动画对象后,通过设置参数 Options 设置动画的属性,相关代码如下。

```
<!-- xxx.hml -->
<div class = "container">
  <div id = "content" class = "box" onclick = "Show"></div>
</div>
/* xxx.css */
.container {
  flex-direction: column;
  justify-content: center;
  align-items: center;
  width: 100%;
}
.box{
  width: 200px;
  height: 200px;
  background-color: #ff0000;
  margin-top: 30px;
```

```
}
/* xxx.js */
export default {
  data: {
    animation: '',
  },
  onInit() {
  },
  onShow() {
    var options = {
      duration: 1500,
      easing: 'ease-in',
      delay: 5,
      iterations: 2,
      direction: 'normal',
    };
    var frames = [
      {
        transform: {
          translate: '-150px -0px'
        }
      },
      {
        transform: {
          translate: '150px 0px'
        }
      }
    ];
    this.animation = this.$element('content').animate(frames, options);
  },
  Show() {
    this.animation.play();
  }
}
```

direction：指定动画的播放模式。normal：动画正向循环播放。reverse：动画反向循环播放。alternate：动画交替循环播放，奇数次正向播放，偶数次反向播放。alternate-reverse：动画反向交替循环播放，奇数次反向播放，偶数次正向播放。

3）添加事件和调用方法

animation 对象支持动画事件和动画方法。可以通过添加开始和取消事件，调用播放、暂停、倒放和结束方法实现预期动画，相关代码请扫描二维码获取。

2. 插值器动画

通过设置插值器实现动画效果，从 API Version 6 开始支持。

（1）创建动画对象。通过 createAnimator 创建一个动画对象，通过设置参数 options 设置动画的属性，相关代码如下。

```html
<!-- xxx.hml -->
<div class = "container">
  <div style = "width: 300px;height: 300px;margin-top: 100px;background: linear-gradient(pink, purple);transform: translate({{translateVal}});">
  </div>
  <div class = "row">
    <button type = "capsule" value = "play" onclick = "playAnimation"></button>
  </div>
</div>
```

```css
/* xxx.css */
.container {
  flex-direction: column;
  align-items: center;
  justify-content: center;
}
button{
  width: 200px;
}
.row{
  width: 65%;
  height: 100px;
  align-items: center;
  justify-content: space-between;
  margin-top: 50px;
  margin-left: 260px;
}
```

```js
/* xxx.js */
import animator from '@ohos.animator';
export default {
  data: {
    translateVal: 0,
    animation: null
  },
  onInit() {},
  onShow(){
    var options = {
      duration: 3000,
      easing:"friction",
      delay:"1000",
      fill: 'forwards',
      direction:'alternate',
      iterations: 2,
      begin: 0,
      end: 180
    }; //设置参数
```

```
      this.animation = animator.createAnimator(options)//创建动画
    },
    playAnimation() {
      var _this = this;
      this.animation.onframe = function(value) {
        _this.translateVal = value
      };
      this.animation.play();
    }
  }
```

使用createAnimator创建动画对象时必须传入options参数。begin为插值起点,不设置时默认为0;end为插值终点,不设置时默认为1。

(2) 添加动画事件和调用接口。animator支持事件和接口,可以通过添加frame、cancel、repeat、finish事件和调用update、play、pause、cancel、reverse、finish接口自定义动画效果,相关代码请扫描二维码获取。

3. 动画帧

本部分包括请求动画帧和取消动画帧。

1) 请求动画帧

通过requestAnimationFrame函数逐帧回调,在调用该函数时传入一个回调函数。runframe在调用requestAnimationFrame时传入带有timestamp参数的回调函数step,将step中的timestamp赋予起始的startTime。当timestamp与startTime的差值小于规定的时间时将再次调用requestAnimationFrame,最终动画将会停止,相关代码如下。

```html
<!-- xxx.hml -->
<div class="container">
  <tabs onchange="changecontent">
    <tab-content>
      <div class="container">
        <stack style="width: 300px; height: 300px; margin-top: 100px; margin-bottom: 100px;">
          <canvas id="mycanvas" style="width: 100%; height: 100%; background-color: coral;">
          </canvas>
          <div style="width: 50px; height: 50px; border-radius: 25px; background-color: indigo; position: absolute; left: {{left}}; top: {{top}};">
          </div>
        </stack>
        <button type="capsule" value="play" onclick="runframe"></button>
      </div>
    </tab-content>
  </tabs>
</div>
/* xxx.css */
```

```css
.container {
  flex-direction: column;
  justify-content: center;
  align-items: center;
  width: 100%;
  height: 100%;
}
button{
  width: 300px;
}
```
```js
/* xxx.js */
export default {
  data: {
    timer: null,
    left: 0,
    top: 0,
    flag: true,
    animation: null,
    startTime: 0,
  },
  onShow() {
    var test = this.$element("mycanvas");
    var ctx = test.getContext("2d");
    ctx.beginPath();
    ctx.moveTo(0, 0);
    ctx.lineTo(300, 300);
    ctx.lineWidth = 5;
    ctx.strokeStyle = "red";
    ctx.stroke();
  },
  runframe() {
    this.left = 0;
    this.top = 0;
    this.flag = true;
    this.animation = requestAnimationFrame(this.step);
  },
  step(timestamp) {
    if (this.flag) {
      this.left += 5;
      this.top += 5;
      if (this.startTime == 0) {
        this.startTime = timestamp;
      }
      var elapsed = timestamp - this.startTime;
      if (elapsed < 500) {
        console.log('callback step timestamp: ' + timestamp);
        this.animation = requestAnimationFrame(this.step);
```

```
        }
      } else {
        this.left -= 5;
        this.top -= 5;
        this.animation = requestAnimationFrame(this.step);
      }
      if (this.left == 250 || this.left == 0) {
        this.flag = !this.flag
      }
    },
    onDestroy() {
      cancelAnimationFrame(this.animation);
    }
}
```

requestAnimationFrame 函数调用回调函数时,在第一个参数位置传入 timestamp 时间戳,表示 requestAnimationFrame 开始执行回调函数的时刻。

2) 取消动画帧

通过 cancelAnimationFrame 函数取消逐帧回调,在调用 cancelAnimationFrame 函数时取消 requestAnimationFrame 函数的请求,相关代码如下。

```
<!-- xxx.hml -->
<div class = "container">
  <tabs onchange = "changecontent">
    <tab-content>
      <div class = "container">
        <stack style = "width: 300px; height: 300px; margin-top: 100px; margin-bottom: 100px;">
          <canvas id = "mycanvas" style = "width: 100%; height: 100%; background-color: coral;">
          </canvas>
          <div style = "width: 50px; height: 50px; border-radius: 25px; background-color: indigo;position: absolute;left: {{left}};top: {{top}};">
          </div>
        </stack>
        <button type = "capsule" value = "play" onclick = "runframe"></button>
      </div>
    </tab-content>
  </tabs>
</div>
/* xxx.css */
.container {
  flex-direction: column;
  justify-content: center;
  align-items: center;
  width: 100%;
```

```css
    height: 100%;
}
button{
    width: 300px;
}
/* xxx.js */
export default {
    data: {
        timer: null,
        left: 0,
        top: 0,
        flag: true,
        animation: null
    },
    onShow() {
        var test = this.$element("mycanvas");
        var ctx = test.getContext("2d");
        ctx.beginPath();
        ctx.moveTo(0, 0);
        ctx.lineTo(300, 300);
        ctx.lineWidth = 5;
        ctx.strokeStyle = "red";
        ctx.stroke();
    },
    runframe() {
        this.left = 0;
        this.top = 0;
        this.flag = true;
        this.animation = requestAnimationFrame(this.step);
    },
    step(timestamp) {
        if (this.flag) {
            this.left += 5;
            this.top += 5;
            this.animation = requestAnimationFrame(this.step);
        } else {
            this.left -= 5;
            this.top -= 5;
            this.animation = requestAnimationFrame(this.step);
        }
        if (this.left == 250 || this.left == 0) {
            this.flag = !this.flag
        }
    },
    onDestroy() {
        cancelAnimationFrame(this.animation);
    }
}
```

}
```

在调用该函数时需传入一个具有标识 ID 的参数。

## 5.6 自定义组件

使用基于 JS 扩展的类 Web 开发范式的方舟开发框架支持自定义组件,用户可根据业务需求将已有的组件进行扩展,增加自定义的私有属性和事件,封装成新的组件,方便在工程中多次调用,提高页面布局代码的可读性,具体封装方法如下。

**1. 构建自定义组件**

相关代码如下。

```
<!-- comp.hml -->
<div class = "item">
    <text class = "title-style">{{title}}</text>
    <text class = "text-style" onclick = "childClicked" focusable = "true">单击这里查看隐藏文本</text>
    <text class = "text-style" if = "{{showObj}}"> hello world </text>
</div>
/* comp.css */
.item {
    width: 700px;
    flex-direction: column;
    height: 300px;
    align-items: center;
    margin-top: 100px;
}
.text-style {
    width: 100%;
    text-align: center;
    font-weight: 500;
    font-family: Courier;
    font-size: 36px;
}
.title-style {
    font-weight: 500;
    font-family: Courier;
    font-size: 50px;
    color: #483d8b;
}
//comp.js
export default {
    props: {
        title: {
            default: 'title',
```
```

```
      },
      showObject: {},
    },
    data() {
      return {
        showObj: this.showObject,
      };
    },
    childClicked () {
      this.$emit('eventType1', {text: '收到子组件参数'});
      this.showObj = !this.showObj;
    },
}
```

2. 引入自定义组件

相关代码如下。

```
<!-- xxx.hml -->
<element name = 'comp' src = '../../common/component/comp.hml'></element>
<div class = "container">
    <text>父组件:{{text}}</text>
    <comp title = "自定义组件" show - object = "{{isShow}}" @event - type1 = "textClicked"></comp>
</div>
/* xxx.css */
.container {
    background - color: #f8f8ff;
    flex: 1;
    flex - direction: column;
    align - content: center;
}
//xxx.js
export default {
    data: {
        text: '开始',
        isShow: false,
    },
    textClicked (e) {
        this.text = e.detail.text;
    },
}
```

本示例中父组件通过添加自定义属性向子组件传递了名称为 title 的参数,子组件在 props 中接收。同时子组件也通过事件绑定向上传递了参数 text,接收时通过 e.detail 获取。如绑定子组件事件,父组件事件命名必须遵循事件绑定规则,自定义组件效果如图 5-10 所示。

图 5-10　自定义组件效果

5.7　JS FA 调用 PA

基于 JS 扩展的类 Web 开发范式的方舟开发框架,提供了 JS FA(Feature Ability)调用 Java PA(Particle Ability)的机制,该机制提供了一种通道传递方法,调用、处理数据返回、上报订阅事件。

当前提供 Ability 和 Internal Ability 两种方式,开发者可以根据业务场景选择合适的调用方式进行开发。

Ability:拥有独立的 Ability 生命周期,FA 使用远端进程通信拉起并请求 PA 服务,适用于基本服务供多 FA 调用或者服务在后台独立运行的场景。

Internal Ability:与 FA 共进程,采用内部函数调用的方式和 FA 进行通信,适用于对服务响应时延要求较高的场景,该方式下 PA 不支持其他 FA 访问调用。

对于 Internal Ability 调用方式的开发,可以使用 js2java-codegen 工具自动生成代码,提高开发效率。

JS 端与 Java 端通过 bundleName 和 abilityName 进行关联。在系统收到 JS 调用请求后,JS 接口中设置的参数选择对应的处理方式。开发者在 onRemoteRequest()中实现 PA 提供的业务逻辑。

1. FA 调用 PA 接口

本部分包括 FA 端和 PA 端提供的接口。

1) FA 端提供以下三个 JS 接口

FeatureAbility.callAbility(OBJECT):调用 PA。

FeatureAbility.subscribeAbilityEvent(OBJECT,Function):订阅 PA。

FeatureAbility.unsubscribeAbilityEvent(OBJECT,Function):取消订阅 PA。

2) PA 端提供以下两类接口

IRemoteObject.onRemoteRequest(int,MessageParcel,MessageParcel,MessageOption):Ability 调用方式,FA 使用远端进程通信拉起并请求 PA 服务。

AceInternalAbility.AceInternalAbilityHandler.onRemoteRequest(int,MessageParcel,MessageParcel,MessageOption):Internal Ability 调用方式,采用内部函数调用的方式和

FA 进行通信。

2．FA 调用 PA 常见问题

callAbility 返回报错：Internal ability not register。返回该错误说明 JS 接口调用请求未在系统中找到对应的 InternalAbilityHandler 进行处理，因此需要检查以下几点是否正确执行。

（1）在 AceAbility 继承类中对 AceInternalAbility 继承类执行了 register 方法。

（2）JS 侧填写的 bundleName 和 abilityName 与 AceInternalAbility 继承类构造函数中填写的名称保持相同，大小写敏感。

（3）检查 JS 端填写的 abilityType（0：Ability；1：Internal Ability），确保没有将 AbilityType 缺省或误填写为 Ability 方式。

Ability 和 Internal Ability 是两种不同的 FA 调用 PA 的方式。Ability 和 InternalAbility 差异项如表 5-2 所示，避免开发时将两者混淆使用。

表 5-2 Ability 和 InternalAbility 差异项

差 异 项	Ability	InternalAbility
JS 端（abilityType）	0	1
是否需要在 config.json 的 abilities 中为 PA 添加声明	需要（有独立的生命周期）	不需要（和 FA 共生命周期）
是否需要在 FA 中注册	不需要	需要
继承的类	ohos.aafwk.ability.Ability	ohos.ace.ability.AceInternalAbility
是否允许被其他 FA 访问调用	是	否

FeatureAbility.callAbility 中 syncOption 参数说明如下。

JS FA 侧返回的结果都是 Promise 对象，因此无论该参数取何值，都采用异步方式等待 PA 侧响应。对于 JAVA PA 侧，在 Internal Ability 方式下收到 FA 的请求后，根据该参数的取值选择通过同步的方式获取结果后返回，或者异步执行 PA 逻辑，获取结果后使用 remoteObject.sendRequest 的方式将结果返回 FA。

使用 await 方式调用时 IDE 编译报错，需引入 babel-runtime/regenerator。

3．示例参考

JS 端调用 FeatureAbility 接口，传入两个 Number 参数，Java 端接收后返回两个数的和。JS FA 应用的 JS 模块（entry/src/main）典型开发目录结构如图 5-11 所示。

1）FA JavaScript 端

使用 Internal Ability 方式时，需要将对应的 action.abilityType 值改为 ABILITY_TYPE_INTERNAL，相关代码如下。

```
//abilityType: 0 - Ability; 1 - Internal Ability
const ABILITY_TYPE_EXTERNAL = 0;
const ABILITY_TYPE_INTERNAL = 1;
//syncOption(Optional, default sync): 0 - Sync; 1 - Async
```

```js
const ACTION_SYNC = 0;
const ACTION_ASYNC = 1;
const ACTION_MESSAGE_CODE_PLUS = 1001;
export default {
  plus: async function() {
    var actionData = {};
    actionData.firstNum = 1024;
    actionData.secondNum = 2048;
    var action = {};
    action.bundleName = 'com.example.hiaceservice';
    action.abilityName = 'com.example.hiaceservice.ComputeServiceAbility';
    action.messageCode = ACTION_MESSAGE_CODE_PLUS;
    action.data = actionData;
    action.abilityType = ABILITY_TYPE_EXTERNAL;
    action.syncOption = ACTION_SYNC;
    var result = await FeatureAbility.callAbility(action);
    var ret = JSON.parse(result);
    if (ret.code == 0) {
        console.info('plus result is:' + JSON.stringify(ret.abilityResult));
    } else {
        console.error('plus error code:' + JSON.stringify(ret.code));
    }
  }
}
```

```
main
├─ config.json
├─ java
│  └─ com.example.hiaceservice.ComputeServiceAbility
│        ├─ ComputeInternalAbility.java
│        ├─ ComputeServiceAbility.java
│        ├─ MainAbility.java
│        ├─ MyApplication.java
│        └─ RequestParam.java
├─ js
│  └─ default
│     ├─ app.js
│     ├─ common （可选）
│     │  ├─ component
│     │  │  └─ component
│     │  │     ├─ componentA.css
│     │  │     ├─ componentA.hml
│     │  │     └─ componentA.js
│     │  └─ images
│     │     ├─ ***.jpg
│     │     └─ ***.png
│     ├─ i18n （可选）
│     │  ├─ en-US.json
│     │  └─ zh-CN.json
│     └─ pages
│        ├─ index
│        │  ├─ index.css
│        │  ├─ index.hml
│        │  └─ index.js
│        └─ detail （可选）
│           ├─ detail.css
│           ├─ detail.hml
│           └─ detail.js
└─ resources （可选）
   ├─ base
   │  ├─ element
   │  │  └─ string.json
   │  └─ media
   │     └─ icon.png
   └─ rawfile
```

图 5-11　JS 模块典型开发目录结构

2) PA 端（Ability 方式）

功能代码实现：在 Java 目录下新建一个 Service Ability，文件命名为 ComputeServiceAbility.java，相关代码如下。

```java
package com.example.hiaceservice;
//ohos 相关接口包
import ohos.aafwk.ability.Ability;
import ohos.aafwk.content.Intent;
import ohos.hiviewdfx.HiLog;
import ohos.hiviewdfx.HiLogLabel;
import ohos.rpc.IRemoteBroker;
import ohos.rpc.IRemoteObject;
import ohos.rpc.RemoteObject;
import ohos.rpc.MessageParcel;
import ohos.rpc.MessageOption;
import ohos.utils.zson.ZSONObject;
```

```java
import java.util.HashMap;
import java.util.Map;
public class ComputeServiceAbility extends Ability {
    //定义日志标签
    private static final HiLogLabel LABEL = new HiLogLabel(HiLog.LOG_APP, 0, "MY_TAG");
    private MyRemote remote = new MyRemote();
    //FA 在请求 PA 服务时会调用 Ability.connectAbility 连接 PA,连接成功后,需要在 onConnect 返
    //回一个 remote 对象,供 FA 向 PA 发送消息
    @Override
    protected IRemoteObject onConnect(Intent intent) {
        super.onConnect(intent);
        return remote.asObject();
    }
    class MyRemote extends RemoteObject implements IRemoteBroker {
        private static final int SUCCESS = 0;
        private static final int ERROR = 1;
        private static final int PLUS = 1001;
        MyRemote() {
            super("MyService_MyRemote");
        }
        @Override
        public boolean onRemoteRequest ( int code, MessageParcel data, MessageParcel reply, MessageOption option) {
            switch (code) {
                case PLUS: {
                    String dataStr = data.readString();
                    RequestParam param = new RequestParam();
                    try {
                        param = ZSONObject.stringToClass(dataStr, RequestParam.class);
                    } catch (RuntimeException e) {
                        HiLog.error(LABEL, "convert failed.");
                    }
                    //返回结果当前仅支持 String,对于复杂结构可以序列化为 Zson 字符串上报
                    Map< String, Object > result = new HashMap< String, Object >();
                    result.put("code", SUCCESS);
                    result.put("abilityResult", param.getFirstNum() + param.getSecondNum());
                    reply.writeString(ZSONObject.toZSONString(result));
                    break;
                }
                default: {
                    Map< String, Object > result = new HashMap< String, Object >();
                    result.put("abilityError", ERROR);
                    reply.writeString(ZSONObject.toZSONString(result));
                    return false;
                }
            }
            return true;
```

```
      }
      @Override
      public IRemoteObject asObject() {
          return this;
      }
  }
}
```

请求参数代码如下。

```
RequestParam.java
public class RequestParam {
    private int firstNum;
    private int secondNum;
    public int getFirstNum() {
        return firstNum;
    }
    public void setFirstNum(int firstNum) {
        this.firstNum = firstNum;
    }
    public int getSecondNum() {
        return secondNum;
    }
    public void setSecondNum(int secondNum) {
        this.secondNum = secondNum;
    }
}
```

3) PA 端（Internal Ability 方式）

功能代码实现可以使用 js2java-codegen 工具自动生成：在 Java 目录下新建一个 Service Ability，文件命名为 ComputeInternalAbility.java，相关代码请扫描二维码获取。

5.8 使用工具自动生成 JS FA 调用 PA 代码

JS FA 调用 PA 是基于 JS 扩展的类 Web 开发范式的方舟开发框架所提供的一种跨语言能力调用的机制，用于建立 JS 能力与 Java 能力之间传递方法调用、处理数据返回及订阅事件上报的通道。开发者可以使用 FA 调用 PA 机制进行应用开发，但直接使用需要手动撰写大量模板代码，且模板代码可能与业务代码相互耦合，使代码可维护性和可读性较差。

为提升开发效率，快速完成 FA 调用 PA 应用，可以在 DevEco Studio 环境中借助 js2java-codegen 工具自动生成 JS FA 调用 PA 代码（目前仅支持 InternalAbility 调用方式）。开发者只需添加简单的配置与标注即可利用该工具完成大部分 FA 调用 PA 模板代码的编写，同时也有效地将业务代码与模板代码相互分离。

1. js2java-codegen 工具简介

js2java-codegen 是工具链提供的自动生成 JS FA 调用 PA 代码的辅助开发工具。它可以根据用户源码生成 FA 调用 PA 所需的、与用户编写的业务代码相互分离的模板代码。

js2java-codegen 工具所支持的 FA 调用 PA 实现方式为 InternalAbility 类型，目前尚不支持 Ability 类型。开发者完成设置后只需编写包含实际业务逻辑的 InternalAbility 类和需要注册的 Ability 类，并在 InternalAbility 类中加上对应注解，js2java-codegen 即可在编译过程中完成 FA 调用 PA 通道的建立。之后，只需在 JS 侧调用由 js2java-codegen 工具生成的 JS 接口即可调用 Java 一侧的能力。

js2java-codegen 工具所生成的模板代码包含 Java 代码和 JS 代码。其中，Java 代码会被直接编译成字节码文件，并且对应 Ability 类中会被自动添加注册与反注册语句，开发者无须关注；而 JS 代码则需要用户手动调用，因此需要在编译前设置好 JS 代码的输出路径。

注解使用说明如下：js2java-codegen 工具通过注解获取信息并生成开发者所需的代码。因此，用户如果使用该工具辅助开发，则需要了解以下三种用法。

1) @InternalAbility 注解

@InternalAbility 注解为类注解，用于 InternalAbility、包含实际业务代码的类（简称 InternalAbility 类），只支持文件中 public 的顶层类，不支持接口类和注解类，包含一个参数 registerTo，值为需要注册到 Ability 类的全名。示例如下，Service 类是一个 InternalAbility 类，注册到位于 com.example 包中的、名为 Ability 的 Ability 类。

```
@InternalAbility(registerTo = "com.example.Ability")
public class Service{}
```

2) @ExportIgnore 注解

@ExportIgnore 注解为方法注解，用于 InternalAbility 类中的某些方法，表示该方法不暴露给 JS 侧调用，仅对 public 方法有效。示例如下，service 方法不会被暴露给 JS 侧。

```
@ExportIgnore
public int service(int input) {
    return input;
}
```

3) @ContextInject 注解

@ContextInject 注解用于 AbilityContext 上的注解。该类由 HarmonyOS 的 Java API 提供，可通过它获取 API 中提供的信息。

可以借助 abilityContext 对象获取 API 中提供的信息，示例如下。

```
@ContextInject
AbilityContext abilityContext;
```

2. 新建工程

体验工具生成模板代码的功能，可使用 DevEco Studio 新建一个包含 JS 前端的简单手机项目，并用其开发一个简单的 FA 调用 PA 应用。

3. 工具开关与编译设置

快速验证功能可选择修改 entry 模块的 build.gradle，通过 entry 模块进行验证。

编译参数位于 ohos→defaultConfig 中,只需添加如下设置即可。开发者需在此处设置 JS 模板代码生成路径,即 jsOutputDir 对应的值。

```
//在文件头部定义 JS 模板代码生成路径
def jsOutputDir = project.file("src/main/js/default/generated").toString()
//在 ohos→defaultConfig 中设置 JS 模板代码生成路径
javaCompileOptions {
    annotationProcessorOptions {
        arguments = ["jsOutputDir": jsOutputDir] //JS 模板代码生成赋值
    }
}
```

工具开关位于 ohos 中,只需添加如下设置即可。值设为 true 则启用工具,值设为 false 则不进行配置,不启用工具。

```
compileOptions {
    f2pautogenEnabled true //此处为启用 js2java-codegen 工具的开关
}
```

4. Java 侧代码编写

模板代码的生成需要提供用于 FA 调用的 PA,因此需要编写 InternalAbility 类,在类上加@InternalAbility 注解,registerTo 参数设为将要注册到的 Ability 类的全称。Ability 类可使用项目中已有的 MainAbility 类,或创建新的 Ability 类。

注意,InternalAbility 类中需要暴露给 FA 调用的方法只能是 public 类型的非静态非 void 方法,若不是则不会被暴露。

一个简单的 InternalAbility 类实现如下,文件名为 Service.java,与 MainAbility 类同包,用注解注册到 MainAbility 类。类中包含一个 ADD 方法作为暴露给 JS FA 调用的能力,实现两数相加的功能,入参为两个 int 参数,返回值为两数之和,示例如下。

```
package com.example.myapplication;
import ohos.annotation.f2pautogen.InternalAbility;
@InternalAbility(registerTo = "com.example.myapplication.MainAbility")
//此处 registerTo 的参数为项目中 MainAbility 类的全称
public class Service {
    public int add(int num1, int num2) {
        return num1 + num2;
    }
}
```

5. 编译

js2java-codegen 工具在编译过程中会自动被调用、生成模板代码并完成整个通道建立的过程。

单击菜单栏中的 Build→Build HAP(s)/APP(s)→Build HAP(s),即可完成项目编译,同时 js2java-codegen 工具会在编译过程中完成 FA 调用 PA 通道的建立。

编译过程会生成Java和JS的模板代码。其中，JS的模板代码位于编译设置中的路径，名称与InternalAbility类相对应；而Java的模板代码位于 entry→build→generated→source→annotation→debug→InternalAbility类同名包→InternalAbility类名＋Stub.java，该类的调用语句会被注入MainAbility类的字节码当中，生成的模板如图5-12和图5-13所示。

```java
package com.example.myapplication;

import ...

public class ServiceStub extends AceInternalAbility {
  public static final String BUNDLE_NAME = "com.example.myapplication";

  public static final String ABILITY_NAME = "com.example.myapplication.MainAbility";

  public static final int ERROR = -1;

  public static final int SUCCESS = 0;

  public static final int OPCODE_add = 0;

  private static ServiceStub instance;

  private Service service;

  private AbilityContext abilityContext;

  public ServiceStub() { super(BUNDLE_NAME, ABILITY_NAME); }
```

图 5-12　Java模板代码示例

```java
package com.example.myapplication;

import ...

public class ServiceStub extends AceInternalAbility {
  public static final String BUNDLE_NAME = "com.example.myapplication";

  public static final String ABILITY_NAME = "com.example.myapplication.MainAbility";

  public static final int ERROR = -1;

  public static final int SUCCESS = 0;

  public static final int OPCODE_add = 0;

  private static ServiceStub instance;

  private Service service;

  private AbilityContext abilityContext;

  public ServiceStub() { super(BUNDLE_NAME, ABILITY_NAME); }
```

图 5-13　JS模板代码示例

6. JS 侧代码编写

为了简易直观地检验工具生成代码的可用性，开发者可以通过修改 entry→src→main→js→default→pages→index→index.js 调用 Java 侧的能力，并在前端页面展示效果。

可通过 import 方式引入 JS 侧 FA 接口，例如 import Service from '../../generated/Service.js';（from 后的值需要与编译设置中的路径进行统一，生成的 JS 代码文件名及类名与 InternalAbility 类名相同）。

一个简单的 index.js 页面实现如下，调用 JS 侧接口，传入 1 和 10 两个参数，并把返回的结果打印在 title 中，这样只要运行该应用就可以验证 FA 调用 PA 是否成功。

```javascript
import Service from '../../generated/Service.js';  //此处 FA 路径和类名对应之前的 jsOutput 路径及 InternalAbility 的名称
export default {
    data: {
        title: "Result:"
    },
    onInit() {
        const echo = new Service(); //此处新建 FA 实例
        echo.add(1,10)
            .then((data) => {
                this.title += data["abilityResult"];   //获取运算结果,并加到 title 之后
            });
    }
}
```

为了方便结果展示，这里对同目录下的 index.hml 也做一些修改，使页面中只显示 title 的内容。

```html
<div class = "container">
    <text class = "title">
        {{ title }}
    </text>
</div>
```

7. 结果验证

启动手机模拟器，成功后运行，看到显示界面则说明 js2java-codegen 工具生成了有效的模板代码，建立了 FA 调用 PA 的通道。

第 6 章 方舟开发框架（ArkUI）——基于 TS 扩展的声明式开发范式

基于 TS 扩展的声明式开发范式的方舟开发框架是为 HarmonyOS 平台开发高性能、跨设备应用设计研发的 UI 开发框架，支持开发者高效构建跨设备应用 UI。

6.1 基于 TS 扩展的声明式开发范式概述

基于 TS 扩展的声明式开发范式的方舟开发框架，采用更接近自然语义的编程方式，可以直观地描述 UI，不必关心框架如何实现 UI 绘制和渲染，从组件、动效和状态管理三个维度提供 UI 能力和系统能力接口，实现系统能力的极简调用，基本功能如下。

开箱即用的组件：框架提供丰富的系统预置组件，可以通过链式调用的方式设置系统组件的渲染效果。开发者可以组合系统组件为自定义组件，通过这种方式将页面组件转换为一个个独立的 UI 单元，实现页面不同单元的独立创建、开发和复用，使页面具有更强的工程性。

丰富的动效接口：提供 svg 标准的绘制图形能力，同时开放了丰富的动效接口，通过封装的物理模型或者调用动画能力接口实现自定义动画轨迹。

状态与数据管理：基于 TS 扩展的声明式开发范式，通过功能不同的装饰器给开发者提供清晰的页面更新渲染流程和管道。状态包括 UI 组件状态和应用程序状态。

系统能力接口：使用基于 TS 扩展的声明式开发范式的方舟开发框架，还封装了丰富的系统能力接口，通过简单的接口调用，实现从 UI 设计到系统能力调用的极简开发。

方舟开发整体框架如图 6-1 所示，主要功能如下。

声明式 UI 前端：UI 开发范式的基础语言规范和多种状态管理机制，内置的 UI 组件、布局和动画，为应用者提供一系列接口支持。

语言运行选择：选用方舟语言运行时，提供了针对 UI 范式语法的解析能力、跨语言调用支持和 TS 语言高性能运行环境。

声明式 UI 后端引擎：后端引擎提供了兼容不同开发范式的 UI 渲染管线、状态管理、绘制能力，以及多种基础组件、布局、动效、交互事件。

渲染引擎：高效的绘制能力，可将渲染管线收集的渲染指令绘制到屏幕。

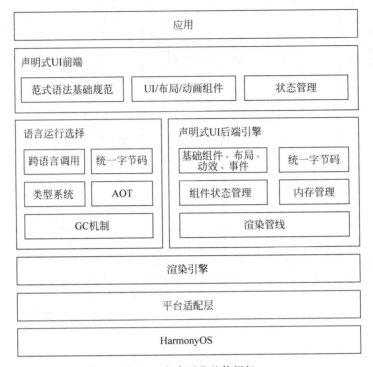

图 6-1 方舟开发整体框架

平台适配层：对系统平台的抽象接口，具备接入不同系统的能力，例如系统渲染管线、生命周期调度等。

6.2 体验声明式 UI

本节将针对具体的开发场景，基于 TS 扩展的声明式开发范式的步骤进行说明，从以下三部分介绍构建健康饮食应用，主要功能为查看食物的营养信息。

Food Detail：构建食物基本信息页面。

Food Category/List：构建食物列表和分类展示页面。

Splash Screen：构建健康饮食应用的闪屏动画。

工程实例参考地址为：https://gitee.com/harmonyos/harmonyos_app_samples/tree/master/ETSUI/eTSBuildCommonView，示例为构建简单页面展示食物番茄的图片和营养信息，主要为了展示简单页面的 Stack 和 Flex 布局。

6.2.1 创建声明式 UI 工程

创建工程之前，需要安装 DevEco Studio，步骤如下。

（1）打开 DevEco Studio，单击 Create Project。如果已有一个工程，则单击 File→New→

New project。

(2) 进入选择 ability template 界面,选择 Empty Ability。

(3) 安装 HarmonyOS sdk。

(4) 进入配置工程界面,将工程名称改为 HealthyDiet,Project Type 选择 Application,Device Type 选择 Phone,Language 选择 eTS,选择兼容 API Version 7。DevEco Studio 会默认将工程保存在 C 盘,如更改工程保存位置,单击 Save Location 的文件夹图标,自行指定工程创建位置,配置完成后单击 Finish 按钮。

(5) 工程创建完成后,打开 app.ets。app.ets 提供应用生命周期的接口:onCreate 和 onDestroy,分别在应用创建之初和应用被销毁时调用。在 app.ets 里可以声明全局变量,声明的数据和方法是整个应用共享的,示例如下。

```
export default {
    onCreate() {
        console.info('Application onCreate')
    },
    onDestroy() {
        console.info('Application onDestroy')
    },
}
```

(6) 在工程导航栏里,打开 index.ets。页面展示了当前的 UI 描述,声明式 UI 框架会自动生成一个组件化的 struct,struct 遵循 Builder 接口声明,在 Build 方法中声明当前的布局和组件,示例如下。

```
@Entry
@Component
struct MyComponent  {
  build() {
    Flex({ direction: FlexDirection.Column, alignItems: ItemAlign.Center, justifyContent: FlexAlign.Center }) {
      Text('Hello World')
        .fontSize(50)
        .fontWeight(FontWeight.Bold)
    }
    .width('100 % ')
    .height('100 % ')
  }
}
```

(7) 单击右侧的 Previewer 按钮,打开预览窗口。可以看到在手机设备类型的预览窗口中 Hello World 居中加粗显示。如果没有 Previewer 按钮,单击 settings→SDK Manager→HarmonyOS Legacy SDK→Tools 查看是否已经安装 Previewer。

(8) 应用安装到手机上运行。将手机连接计算机,待 IDE 识别到物理设备后,单击

entry 按钮。

在安装之前,需要配置应用签名。安装成功后,单击屏幕上的 Run 图标打开应用,可以看到居中加粗显示的 Hello World。

6.2.2 初识 Component

在自定义组件之前,需要先了解什么是组件和装饰器,并进行组件初始化,然后通过修改组件属性和构造参数,实现一个自定义组件。

1. 组件和装饰器

在声明式 UI 中,所有的页面都由组件构成。组件的数据结构为 struct,装饰器 @Component 是组件化的标志。用 @Component 修饰的 struct 表示此结构体有了组件化的能力。

自定义组件的声明方式如下。

```
@Component
struct MyComponent {}
```

在 IDE 创建工程模板中,MyComponent 可以居中显示文字的自定义组件。开发者可以在 Component 的 Build 方法中描述自己的 UI 结构,但需要遵循 Builder 的接口约束。

```
interface Builder {
    build: () => void
}
```

@Entry 修饰的 Component 表示 Component 是页面的总入口,也可以理解为页面的根节点。一个页面有且仅能有一个 @Entry,只有被 @Entry 修饰的组件或者其子组件,才会在页面上显示。

@Component 和 @Entry 都是基础且十分重要的装饰器,装饰器就是某一种修饰,给被装饰的对象赋予某种能力,例如 @Entry 是页面入口的能力,@Component 是组件化能力。

2. 修改组件属性和构造参数

创建系统组件时,会显示其默认样式,可以通过更改组件的属性样式改变组件的视图显示。

修改 Text 组件的 fontSize 属性更改组件的字体大小,将字体大小设置为 26,修改 fontWeight 属性设置字体粗细为 500。number 类型的取值范围为 100～900,默认为 400,取值越大,字体越粗。fontWeight 为内置枚举类型,取值支持 Lighter、Normal、Bold 和 Bolder。

属性方法要紧随组件,通过"."运算符连接,也可以通过链式调用的方式配置组件的多个属性,示例如下。

```
@Entry
@Component
```

```
struct MyComponent {
    build() {
        Flex({ direction: FlexDirection.Column, alignItems: ItemAlign.Center, justifyContent: FlexAlign.Center }) {
            Text('Hello World')
                .fontSize(26)
                .fontWeight(500)
        }
        .width('100%')
        .height('100%')
    }
}
```

修改 Text 组件的显示内容 Hello World 为 Tomato,修改 Text 组件的构造参数相关代码如下。

```
@Entry
@Component
struct MyComponent {
    build() {
        Flex({ direction: FlexDirection.Column, alignItems: ItemAlign.Center, justifyContent: FlexAlign.Center }) {
            Text('Tomato')
                .fontSize(26)
                .fontWeight(500)
        }
        .width('100%')
        .height('100%')
    }
}
```

6.2.3 创建简单视图

本节介绍如何通过容器组件 Stack、Flex 和基本组件 Image、Text,构建用户自定义组件,完成图文并茂的食物介绍。

1. 构建 Stack 布局

构建 Stack 布局方法如下。

(1) 创建食物名称。删除工程模板 Build 方法的代码,创建 Stack 组件,将 Text 组件放进 Stack 组件的花括号中,使其成为 Stack 组件的子组件。Stack 组件为堆叠组件,可以包含一个或多个子组件,其特点是后一个子组件覆盖前一个子组件,相关代码如下。

```
@Entry
@Component
struct MyComponent {
    build() {
```

```
        Stack() {
            Text('Tomato')
                .fontSize(26)
                .fontWeight(500)
        }
    }
}
```

（2）食物图片展示。创建 Image 组件，指定 Image 组件的 URL，Text 和 Image 组件都是必选构造参数组件。为了让 Text 在 Image 组件上方显示，需要先声明 Image 组件。图片资源放在 resources 下的 rawfile 文件夹内，引用 rawfile 下资源时使用"$rawfile('filename')"的形式，filename 为 rawfile 目录下的文件相对路径。当前 $rawfile 仅支持 Image 控件引用图片资源，相关代码如下。

```
@Entry
@Component
struct MyComponent {
    build() {
        Stack() {
            Image( $rawfile('Tomato.png'))
            Text('Tomato')
                .fontSize(26)
                .fontWeight(500)
        }
    }
}
```

（3）通过资源访问图片。除指定图片路径外，也可以引用媒体资源符 $r 引用资源，需要遵循 resources 文件夹的资源限定词的规则。右击 resources 文件夹，单击 New → Resource Directory，选择 Resource Type 为 Media(图片资源)，选择资源限定词为 Device-Phone(目前开发设备为手机)。单击 OK 按钮后，resources 文件夹下生成 phone.media 文件夹，将 Tomato.png 放入该文件夹内，可以通过"$r('app.type.name')"的形式引用应用资源，即 $r('app.media.Tomato')，相关代码如下。

```
@Entry
@Component
struct MyComponent {
    build() {
        Stack() {
            Image( $r('app.media.Tomato'))
                .objectFit(ImageFit.Contain)
                .height(357)
            Text('Tomato')
                .fontSize(26)
                .fontWeight(500)
```

        }
      }
    }

(4) 设置 Image 宽高,并将 image 的 objectFit 属性设置为 ImageFit.Contain,即保持图片长宽比的情况下,使图片完整地显示在边界内。

Image 填满整个屏幕原因如下:Image 没有设置宽高。Image 的 objectFit 默认属性是 ImageFit.Cover,在保持长宽比的情况下放大或缩小,使其填满整个显示边界,相关代码如下。

```
@Entry
@Component
struct MyComponent {
  build() {
    Stack() {
        Image( $r('app.media.Tomato'))
            .objectFit(ImageFit.Contain)
            .height(357)
        Text('Tomato')
            .fontSize(26)
            .fontWeight(500)
    }
  }
}
```

(5) 设置食物图片和名称布局。设置 Stack 的对齐方式为底部起始端对齐,Stack 默认为居中对齐。设置 Stack 构造参数 alignContent 为 Alignment.BottomStart。其中 Alignment 和 FontWeight 一样,都是框架提供的内置枚举类型,相关代码如下。

```
@Entry
@Component
struct MyComponent {
  build() {
    Stack({ alignContent: Alignment.BottomStart }) {
        Image( $r('app.media.Tomato'))
            .objectFit(ImageFit.Contain)
            .height(357)
        Text('Tomato')
            .fontSize(26)
            .fontWeight(500)
    }
  }
}
```

(6) 设置 Stack 的背景颜色改变食物图片的背景颜色,方法如下。

通过框架提供的 Color 内置枚举值设置,例如 backgroundColor(Color.Red),即设置背景颜色为红色。

string 类型参数支持的颜色格式如下：rgb、rgba 和 HEX 颜色码。例如 backgroundColor('#0000FF')，即设置背景颜色为蓝色，backgroundColor('rgb(255,255,255)')，即设置背景颜色为白色，相关代码如下。

```
@Entry
@Component
struct MyComponent {
  build() {
    Stack({ alignContent: Alignment.BottomStart }) {
      Image( $r('app.media.Tomato'))
        .objectFit(ImageFit.Contain)
        .height(357)
      Text('Tomato')
        .fontSize(26)
        .fontWeight(500)
    }
    .backgroundColor('#FFedf2f5')
  }
}
```

(7) 调整 Text 组件的外边距 margin，使其左侧和底部有一定的距离。margin 是简写属性，可以统一指定四个边的外边距，也可以分别指定，具体设置方式如下。

参数为 Length 时，统一指定四个边的外边距，例如 margin(20)，上、右、下、左四个边的外边距都是 20。

参数为{top?: Length, right?: Length, bottom?: Length, left?: Length}，分别指定四个边的边距，例如 margin({ left: 26, bottom: 17.4 })，左边距为 26，下边距为 17.4，相关代码如下。

```
@Entry
@Component
struct MyComponent {
  build() {
    Stack({ alignContent: Alignment.BottomStart }) {
      Image( $r('app.media.Tomato'))
        .objectFit(ImageFit.Contain)
        .height(357)
      Text('Tomato')
        .fontSize(26)
        .fontWeight(500)
        .margin({left: 26, bottom: 17.4})
    }
    .backgroundColor('#FFedf2f5')
  }
}
```

(8) 调整组件间的结构，语义化组件名称。创建页面入口组件为 FoodDetail，在 FoodDetail 中创建 Column，设置水平方向上居中对齐 alignItems(HorizontalAlign.Center)。MyComponent 组件名改为 FoodImageDisplay，为 FoodDetail 的子组件。

Column 是子组件竖直排列的容器组件，本质为线性布局，所以只能设置交叉轴方向的对齐，相关代码如下。

```
@Component
struct FoodImageDisplay {
  build() {
    Stack({ alignContent: Alignment.BottomStart }) {
      Image( $r('app.media.Tomato'))
        .objectFit(ImageFit.Contain)
      Text('Tomato')
        .fontSize(26)
        .fontWeight(500)
        .margin({ left: 26, bottom: 17.4 })
    }
    .height(357)
    .backgroundColor('#FFedf2f5')
  }
}
@Entry
@Component
struct FoodDetail {
  build() {
    Column() {
      FoodImageDisplay()
    }
    .alignItems(HorizontalAlign.Center)
  }
}
```

2. 构建 Flex 布局

使用 Flex 弹性布局构建食物成分表，弹性布局在本场景的优势是可以免去多余的宽高计算，通过比例设置不同单元格的大小。

(1) 创建 ContentTable 组件，使其成为页面入口组件 FoodDetail 的子组件，相关代码如下。

```
@Component
struct FoodImageDisplay {
  build() {
    Stack({ alignContent: Alignment.BottomStart }) {
      Image( $r('app.media.Tomato'))
        .objectFit(ImageFit.Contain)
        .height(357)
```

```
        Text('Tomato')
          .fontSize(26)
          .fontWeight(500)
          .margin({ left: 26, bottom: 17.4 })
      }
      .backgroundColor('#FFedf2f5')
    }
  }
  @Component
  struct ContentTable {
    build() {}
  }
  @Entry
  @Component
  struct FoodDetail {
    build() {
      Column() {
        FoodImageDisplay()
        ContentTable()
      }
      .alignItems(HorizontalAlign.Center)
    }
  }
```

(2) 创建 Flex 组件展示 Tomato 两类成分。一类是热量 Calories，包含卡路里 (Calories)；另一类是营养成分 Nutrition，包含蛋白质 (Protein)、脂肪 (Fat)、碳水化合物 (Carbohydrates) 和维生素 C(VitaminC)。

创建热量。创建 Flex 组件，高度为 280，上、右、左内边距为 30，3 个 Text 子组件分别代表类别名 (Calories)、含量名称 (Calories) 和含量数值 (17kcal)，Flex 组件默认为水平排列方式。

ContentTable 扩展代码如下。

```
@Component
struct ContentTable {
  build() {
    Flex() {
      Text('Calories')
        .fontSize(17.4)
        .fontWeight(FontWeight.Bold)
      Text('Calories')
        .fontSize(17.4)
      Text('17kcal')
        .fontSize(17.4)
    }
    .height(280)
```

```
      .padding({ top: 30, right: 30, left: 30 })
  }
}
@Entry
@Component
struct FoodDetail {
  build() {
    Column() {
      FoodImageDisplay()
      ContentTable()
    }
    .alignItems(HorizontalAlign.Center)
  }
}
```

(3) 调整布局,设置各部分占比。分类名占比(layoutWeight)为1,成分名和成分含量共占比(layoutWeight)为2。成分名和成分含量位于同一个Flex中,成分名占据所有剩余空间flexGrow(1),相关代码如下。

```
@Component
struct FoodImageDisplay {
  build() {
    Stack({ alignContent: Alignment.BottomStart }) {
      Image( $m('Tomato.png'))
        .objectFit(ImageFit.Contain)
        .height(357)
      Text('Tomato')
        .fontSize(26)
        .fontWeight(500)
        .margin({ left: 26, bottom: 17.4 })
    }
    .backgroundColor('#FFedf2f5')
  }
}
@Component
struct ContentTable {
  build() {
    Flex() {
      Text('Calories')
        .fontSize(17.4)
        .fontWeight(FontWeight.Bold)
        .layoutWeight(1)
      Flex() {
        Text('Calories')
          .fontSize(17.4)
          .flexGrow(1)
        Text('17kcal')
```

```
          .fontSize(17.4)
        }
        .layoutWeight(2)
      }
      .height(280)
      .padding({ top: 30, right: 30, left: 30 })
    }
  }
}
@Entry
@Component
struct FoodDetail {
  build() {
    Column() {
      FoodImageDisplay()
      ContentTable()
    }
    .alignItems(HorizontalAlign.Center)
  }
}
```

(4) 仿照热量分类创建营养成分分类。营养成分包含蛋白质、脂肪、碳水化合物和维生素 C 共 4 类,后 3 类在表格中省略分类名,用空格代替。

设置外层 Flex 为竖直排列 FlexDirection.Column,在主轴方向(竖直方向)上等距排列 FlexAlign.SpaceBetween,在交叉轴方向(水平轴方向)上首部对齐排列 ItemAlign.Start。相关代码如下。

```
@Component
struct ContentTable {
  build() {
    Flex ({ direction: FlexDirection.Column, justifyContent: FlexAlign.SpaceBetween,
      alignItems: ItemAlign.Start }) {
      Flex() {
        Text('Calories')
          .fontSize(17.4)
          .fontWeight(FontWeight.Bold)
          .layoutWeight(1)
        Flex() {
          Text('Calories')
            .fontSize(17.4)
            .flexGrow(1)
          Text('17kcal')
            .fontSize(17.4)
        }
        .layoutWeight(2)
      }
      Flex() {
```

```
Text('Nutrition')
  .fontSize(17.4)
  .fontWeight(FontWeight.Bold)
  .layoutWeight(1)
Flex() {
  Text('Protein')
    .fontSize(17.4)
    .flexGrow(1)
  Text('0.9g')
    .fontSize(17.4)
}
.layoutWeight(2)
}
Flex() {
  Text('')
    .fontSize(17.4)
    .fontWeight(FontWeight.Bold)
    .layoutWeight(1)
  Flex() {
    Text('Fat')
      .fontSize(17.4)
      .flexGrow(1)
    Text('0.2g')
      .fontSize(17.4)
  }
  .layoutWeight(2)
}
Flex() {
  Text('')
    .fontSize(17.4)
    .fontWeight(FontWeight.Bold)
    .layoutWeight(1)
  Flex() {
    Text('Carbohydrates')
      .fontSize(17.4)
      .flexGrow(1)
    Text('3.9g')
      .fontSize(17.4)
  }
  .layoutWeight(2)
}
Flex() {
  Text('')
    .fontSize(17.4)
    .fontWeight(FontWeight.Bold)
    .layoutWeight(1)
  Flex() {
```

```
                Text('vitaminC')
                    .fontSize(17.4)
                    .flexGrow(1)
                Text('17.8mg')
                    .fontSize(17.4)
            }
            .layoutWeight(2)
        }
    }
    .height(280)
    .padding({ top: 30, right: 30, left: 30 })
  }
}
@Entry
@Component
struct FoodDetail {
    build() {
        Column() {
            FoodImageDisplay()
            ContentTable()
        }
        .alignItems(HorizontalAlign.Center)
    }
}
```

（5）使用自定义构造函数@Builder简化代码。可以发现，每个成分表中的单元其实都是一样的UI结构。

当前对每个成分单元都进行了声明，造成了代码的重复和冗余。可以使用@Builder构建自定义方法，抽象出相同的UI结构声明。@Builder修饰方法和Component的Build方法都是为了声明一些UI渲染结构，遵循一样的eTS语法。开发者可以定义一个或者多个@Builder修饰方法，但Component的build方法只有一个。

在ContentTable内声明@Builder修饰的IngredientItem方法，用于声明分类名、成分名称和成分含量UI描述，相关代码如下。

```
@Builder IngredientItem(title:string, name: string, value: string) {
    Flex() {
        Text(title)
            .fontSize(17.4)
            .fontWeight(FontWeight.Bold)
            .layoutWeight(1)
        Flex() {
            Text(name)
                .fontSize(17.4)
                .flexGrow(1)
            Text(value)
```

```
      .fontSize(17.4)
    }
    .layoutWeight(2)
  }
}
```

在 ContentTable 的 build 方法内调用 IngredientItem 接口,需要 This 调用 Component 作用域内的方法,以此区分全局的方法调用,相关代码如下。

```
@Component
struct ContentTable {
  ......
  build() {
    Flex({ direction: FlexDirection.Column, justifyContent: FlexAlign.SpaceBetween, alignItems: ItemAlign.Start }) {
      this.IngredientItem('Calories', 'Calories', '17kcal')
      this.IngredientItem('Nutrition', 'Protein', '0.9g')
      this.IngredientItem('', 'Fat', '0.2g')
      this.IngredientItem('', 'Carbohydrates', '3.9g')
      this.IngredientItem('', 'VitaminC', '17.8mg')
    }
    .height(280)
    .padding({ top: 30, right: 30, left: 30 })
  }
}
```

ContentTable 组件整体代码如下。

```
@Component
struct ContentTable {
  @Builder IngredientItem(title:string, name: string, value: string) {
    Flex() {
      Text(title)
        .fontSize(17.4)
        .fontWeight(FontWeight.Bold)
        .layoutWeight(1)
      Flex() {
        Text(name)
          .fontSize(17.4)
          .flexGrow(1)
        Text(value)
          .fontSize(17.4)
      }
      .layoutWeight(2)
    }
  }
  build() {
```

```
        Flex({ direction: FlexDirection.Column, justifyContent: FlexAlign.SpaceBetween, alignItems:
ItemAlign.Start }) {
            this.IngredientItem('Calories', 'Calories', '17kcal')
            this.IngredientItem('Nutrition', 'Protein', '0.9g')
            this.IngredientItem('', 'Fat', '0.2g')
            this.IngredientItem('', 'Carbohydrates', '3.9g')
            this.IngredientItem('', 'VitaminC', '17.8mg')
        }
        .height(280)
        .padding({ top: 30, right: 30, left: 30 })
    }
}
@Entry
@Component
struct FoodDetail {
    build() {
        Column() {
            FoodImageDisplay()
            ContentTable()
        }
        .alignItems(HorizontalAlign.Center)
    }
}
```

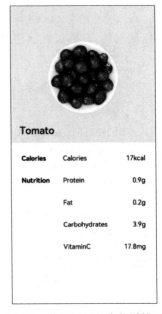

通过学习 Stack、Flex 布局已完成食物的图文展示和营养成分表，构建出第一个普通视图的食物详情页，如图 6-2 所示。

图 6-2　普通视图的食物详情页

6.3　页面布局与连接

本节主要介绍构建食物分类列表页面和食物详情页，展示 List、Grid 布局及页面路由的基本用法。其他用法可参考 eTSDefiningPageLayoutAndConnection 示例，链接地址为：https://gitee.com/harmonyos/harmonyos_app_samples/tree/master/ETSUI/eTSDefiningPageLayoutAndConnection。

6.3.1　构建数据模型

在创建视图中，逐一表述食物的各个信息，如食物名称、卡路里、蛋白质、脂肪、碳水和维生素 C。这样的编码形式在实际的开发中是不切实际的，所以要创建食物数据模型统一存储和管理数据，步骤如下。

（1）新建 model 文件夹，在 model 目录下创建 FoodData.ets，如图 6-3 所示。

（2）定义食物数据的存储模型 FoodData 和枚举变量 Category，FoodData 类包含食物 ID、名称、分类、图片、热量、蛋白质、脂肪、碳水和维生素 C 属性。eTS 语言是在 TS 语言基

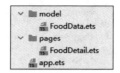

图 6-3　新建 model 文件夹

础上的扩展,同样支持 TS 语法,相关代码如下。

```
enum Category {
  Fruit,
  Vegetable,
  Nut,
  Seafood,
  Dessert
}
let NextId = 0;
class FoodData {
  id: string;
  name: string;
  image: Resource;
  category: Category;
  calories: number;
  protein: number;
  fat: number;
  carbohydrates: number;
  vitaminC: number;
  constructor(name: string, image: Resource, category: Category, calories: number, protein: number, fat: number, carbohydrates: number, vitaminC: number) {
    this.id = `${ NextId++ }`;
    this.name = name;
    this.image = image;
    this.category = category;
    this.calories = calories;
    this.protein = protein;
    this.fat = fat;
    this.carbohydrates = carbohydrates;
    this.vitaminC = vitaminC;
  }
}
```

(3) 在 resources→phone→media 目录下存入食物图片资源,图片名称为食物名称,如图 6-4 所示。

(4) 创建食物资源数据。在 model 文件夹下创建 FoodDataModels.ets,声明食物成分数组 FoodComposition。以 12 个食物数据为例,实际开发中,可以自定义更多的数据资源,当食物资源很多时,建议使用数据懒加载 LazyForEach,以下营养数据均来自网络。

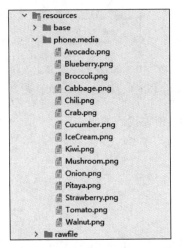

图 6-4 存入食物图片资源

```
const FoodComposition: any[ ] = [
    { 'name': 'Tomato', 'image': $r('app.media.Tomato'), 'category': Category.Vegetable, 'calories': 17, 'protein': 0.9, 'fat': 0.2, 'carbohydrates': 3.9, 'vitaminC': 17.8 },
    { 'name': 'Walnut', 'image': $r('app.media.Walnut'), 'category': Category.Nut, 'calories': 654, 'protein': 15, 'fat': 65, 'carbohydrates': 14, 'vitaminC': 1.3 },
    { 'name': 'Cucumber', 'image': $r('app.media.Cucumber'), 'category': Category.Vegetable, 'calories': 30, 'protein': 3, 'fat': 0, 'carbohydrates': 1.9, 'vitaminC': 2.1 },
    { 'name': 'Blueberry', 'image': $r('app.media.Blueberry'), 'category': Category.Fruit, 'calories': 57, 'protein': 0.7, 'fat': 0.3, 'carbohydrates': 14, 'vitaminC': 9.7 },
    { 'name': 'Crab', 'image': $r('app.media.Crab'), 'category': Category.Seafood, 'calories': 97, 'protein': 19, 'fat': 1.5, 'carbohydrates': 0, 'vitaminC': 7.6 },
    { 'name': 'IceCream', 'image': $r('app.media.IceCream'), 'category': Category.Dessert, 'calories': 207, 'protein': 3.5, 'fat': 11, 'carbohydrates': 24, 'vitaminC': 0.6 },
    { 'name': 'Onion', 'image': $r('app.media.Onion'), 'category': Category.Vegetable, 'calories': 39, 'protein': 1.1, 'fat': 0.1, 'carbohydrates': 9, 'vitaminC': 7.4 },
    { 'name': 'Mushroom', 'image': $r('app.media.Mushroom'), 'category': Category.Vegetable, 'calories': 22, 'protein': 3.1, 'fat': 0.3, 'carbohydrates': 3.3, 'vitaminC': 2.1 },
    { 'name': 'Kiwi', 'image': $r('app.media.Kiwi'), 'category': Category.Fruit, 'calories': 60, 'protein': 1.1, 'fat': 0.5, 'carbohydrates': 15, 'vitaminC': 20.5 },
    { 'name': 'Pitaya', 'image': $r('app.media.Pitaya'), 'category': Category.Fruit, 'calories': 60, 'protein': 1.2, 'fat': 0, 'carbohydrates': 10, 'vitaminC': 60.9 },
    { 'name': 'Avocado', 'image': $r('app.media.Avocado'), 'category': Category.Fruit, 'calories': 160, 'protein': 2, 'fat': 15, 'carbohydrates': 9, 'vitaminC': 10 },
    { 'name': 'Strawberry', 'image': $r('app.media.Strawberry'), 'category': Category.Fruit, 'calories': 32, 'protein': 0.7, 'fat': 0.3, 'carbohydrates': 8, 'vitaminC': 58.8 }
]
```

(5) 创建 initializeOnStartUp 方法初始化 FoodData 的数组。在 FoodDataModels.ets 中使用定义 FoodData.ets 的 FoodData 和 Category，所以要将 FoodData.ets 的 FoodData

类 export，在 FoodDataModels.ets 内 import FoodData 和 Category，相关代码如下。

```
//FoodData.ets
export enum Category {
......
}
export class FoodData {
  ......
}
//FoodDataModels.ets
import { Category, FoodData } from './FoodData'
export function initializeOnStartup(): Array<FoodData> {
  let FoodDataArray: Array<FoodData> = []
  FoodComposition.forEach(item => {
    FoodDataArray.push(new FoodData(item.name, item.image, item.category, item.calories,
item.protein, item.fat, item.carbohydrates, item.vitaminC ));
  })
  return FoodDataArray;
}
```

6.3.2 构建列表 List 布局

使用 List 组件和 ForEach 循环渲染，构建食物列表布局。

(1) 在 pages 目录新建页面 FoodCategoryList.ets，将 index.ets 改名为 FoodDetail.ets，并将其添加到 config.json 文件下的 pages 标签，位于第一序位的页面为首页，相关代码如下。

```
"js": [
  {
    "pages": [
      "pages/FoodCategoryList",
      "pages/FoodDetail"
    ],
]
```

(2) 新建 FoodList 组件作为页面入口组件，FoodListItem 为其子组件。List 组件是列表组件，适用于重复同类数据的展示，其子组件为 ListItem，适用于展示列表中的单元，相关代码如下。

```
@Component
struct FoodListItem {
  build() {}
}
@Entry
@Component
struct FoodList {
```

```
    build() {
      List() {
        ListItem() {
          FoodListItem()
        }
      }
    }
  }
```

(3) 引入 FoodData 类和 initializeOnStartup 方法。

```
import { FoodData } from '../model/FoodData'
import { initializeOnStartup } from '../model/FoodDataModels'
```

(4) 使用 FoodList 和 FoodListItem 组件完成数值传递。在 FoodList 组件内创建类型为 FoodData[]的成员变量 foodItems，调用 initializeOnStartup 方法为其赋值。在 FoodListItem 组件内创建类型为 FoodData 的成员变量 foodItem。将父组件 foodItems 数组的第一个元素的 foodItems[0]作为参数传递给 FoodListItem，相关代码如下。

```
import { FoodData } from '../model/FoodData'
import { initializeOnStartup } from '../model/FoodDataModels'
@Component
struct FoodListItem {
  private foodItem: FoodData
  build() {}
}
@Entry
@Component
struct FoodList {
  private foodItems: FoodData[] = initializeOnStartup()
  build() {
    List() {
      ListItem() {
        FoodListItem({ foodItem: this.foodItems[0] })
      }
    }
  }
}
```

(5) 声明子组件 FoodListItem 的 UI 布局。创建 Flex 组件，包含食物图片缩略图、食物名称和食物对应的卡路里，相关代码如下。

```
import { FoodData } from '../model/FoodData'
import { initializeOnStartup } from '../model/FoodDataModels'
@Component
struct FoodListItem {
  private foodItem: FoodData
```

```
  build() {
    Flex({ justifyContent: FlexAlign.Start, alignItems: ItemAlign.Center }) {
      Image(this.foodItem.image)
        .objectFit(ImageFit.Contain)
        .height(40)
        .width(40)
        .backgroundColor('#FFf1f3f5')
        .margin({ right: 16 })
      Text(this.foodItem.name)
        .fontSize(14)
        .flexGrow(1)
      Text(this.foodItem.calories + 'kcal')
        .fontSize(14)
    }
    .height(64)
    .margin({ right: 24, left:32 })
  }
}
@Entry
@Component
struct FoodList {
  private foodItems: FoodData[] = initializeOnStartup()
  build() {
    List() {
      ListItem() {
        FoodListItem({ foodItem: this.foodItems[0] })
      }
    }
  }
}
```

(6) 在 List 组件创建两个 FoodListItem，分别给 FoodListItem 传递 foodItems 数组的第一个元素 this.foodItems[0]和第二个元素 foodItem：this.foodItems[1]，相关代码如下。

```
import { FoodData } from '../model/FoodData'
import { initializeOnStartup } from '../model/FoodDataModels'
@Component
struct FoodListItem {
    private foodItem: FoodData
    build() {
        Flex({ justifyContent: FlexAlign.Start, alignItems: ItemAlign.Center }) {
            Image(this.foodItem.image)
                .objectFit(ImageFit.Contain)
                .height(40)
                .width(40)
                .backgroundColor('#FFf1f3f5')
                .margin({ right: 16 })
```

```
            Text(this.foodItem.name)
                .fontSize(14)
                .flexGrow(1)
            Text(this.foodItem.calories + 'kcal')
                .fontSize(14)
        }
        .height(64)
        .margin({ right: 24, left:32 })
    }
}
@Entry
@Component
struct FoodList {
    private foodItems: FoodData[] = initializeOnStartup()
    build() {
        List() {
            ListItem() {
                FoodListItem({ foodItem: this.foodItems[0] })
            }
            ListItem() {
                FoodListItem({ foodItem: this.foodItems[1] })
            }
        }
    }
}
```

(7) 单独创建每个 FoodListItem 是不合理的,需要引入 ForEach 循环渲染,ForEach 语法如下,相关代码如下。

```
ForEach(
    arr: any[], // Array to be iterated
    itemGenerator: (item: any) => void, //child component generator
    keyGenerator?: (item: any) => string //(optional) Unique key generator, which is recommended.
)
```

ForEach 组有三个参数,第一个参数是需要被遍历的数组,第二个参数是生成子组件的 lambda 函数,第三个参数是键值生成器。出于性能原因,使第三个参数是可选的,建议由开发者提供。keyGenerator 使开发框架能够更好地识别数组更改,而不必因为 item 的更改重建全部节点。遍历 foodItems 数组循环创建 ListItem 组件,foodItems 中每个 item 都作为参数传递给 FoodListItem 组件,相关代码如下。

```
ForEach(this.foodItems, item => {
    ListItem() {
        FoodListItem({ foodItem: item })
    }
```

}, item => item.id.toString())
```

**整体代码如下。**

```
import { FoodData } from '../model/FoodData'
import { initializeOnStartup } from '../model/FoodDataModels'
@Component
struct FoodListItem {
  private foodItem: FoodData
  build() {
    Flex({ justifyContent: FlexAlign.Start, alignItems: ItemAlign.Center }) {
      Image(this.foodItem.image)
        .objectFit(ImageFit.Contain)
        .height(40)
        .width(40)
        .backgroundColor('#FFf1f3f5')
        .margin({ right: 16 })
      Text(this.foodItem.name)
        .fontSize(14)
        .flexGrow(1)
      Text(this.foodItem.calories + 'kcal')
        .fontSize(14)
    }
    .height(64)
    .margin({ right: 24, left:32 })
  }
}
@Entry
@Component
struct FoodList {
  private foodItems: FoodData[] = initializeOnStartup()
  build() {
    List() {
      ForEach(this.foodItems, item => {
        ListItem() {
          FoodListItem({ foodItem: item })
        }
      }, item => item.id.toString())
    }
  }
}
```

(8) 添加 FoodList 标题,相关代码如下。

```
@Entry
@Component
struct FoodList {
  private foodItems: FoodData[] = initializeOnStartup()
```
```

```
build() {
  Column() {
    Flex({justifyContent: FlexAlign.Start, alignItems: ItemAlign.Center}) {
      Text('Food List')
        .fontSize(20)
        .margin({ left:20 })
    }
    .height('7%')
    .backgroundColor('#FFf1f3f5')
    List() {
      ForEach(this.foodItems, item => {
        ListItem() {
          FoodListItem({ foodItem: item })
        }
      }, item => item.id.toString())
    }
    .height('93%')
  }
}
```

运行结果如图 6-5 所示。

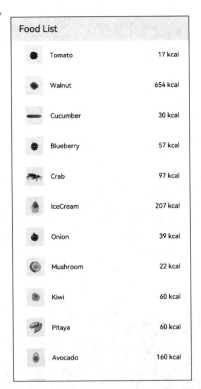

图 6-5　添加 FoodList 标题运行结果

6.3.3 构建分类 Grid 布局

健康饮食应用在主页提供给用户两种食物显示方式：列表显示和网格显示。通过页签切换不同食物分类的网格布局步骤如下。

（1）将 Category 枚举类型引入 FoodCategoryList 页面。

```
import { Category, FoodData } from '../model/FoodData'
```

（2）创建 FoodCategoryList 和 FoodCategory 组件，其中 FoodCategoryList 作为新的页面入口组件，在入口组件调用 initializeOnStartup 方法，相关代码如下。

```
@Component
struct FoodList {
  private foodItems: FoodData[]
  build() {
    ......
  }
}
@Component
struct FoodCategory {
  private foodItems: FoodData[]
  build() {
    ......
  }
}
@Entry
@Component
struct FoodCategoryList {
  private foodItems: FoodData[] = initializeOnStartup()
  build() {
    ......
  }
}
```

（3）在 FoodCategoryList 组件内创建 showList 成员变量，用于控制 List 和 Grid 布局的渲染切换，需要用到条件渲染语句，相关代码如下。

```
@Entry
@Component
struct FoodCategoryList {
  private foodItems: FoodData[] = initializeOnStartup()
  private showList: boolean = false
  build() {
    Stack() {
      if (this.showList) {
        FoodList({ foodItems: this.foodItems })
```

```
        } else {
          FoodCategory({ foodItems: this.foodItems })
        }
      }
    }
  }
```

（4）在页面右上角创建切换 List/Grid 布局的图标。设置 Stack 对齐方式为顶部尾部对齐 TopEnd，创建 Image 组件，设置其单击事件，即 showList 取反，相关代码如下。

```
@Entry
@Component
struct FoodCategoryList {
  private foodItems: FoodData[] = initializeOnStartup()
  private showList: boolean = false
  build() {
    Stack({ alignContent: Alignment.TopEnd }) {
      if (this.showList) {
        FoodList({ foodItems: this.foodItems })
      } else {
        FoodCategory({ foodItems: this.foodItems })
      }
      Image( $r('app.media.Switch'))
        .height(24)
        .width(24)
        .margin({ top: 15, right: 10 })
        .onClick(() => {
          this.showList = !this.showList
        })
    }.height('100%')
  }
}
```

（5）添加@State 装饰器。单击右上角的 switch 标签后，页面没有任何变化，这是因为 showList 不是有状态数据，它的变化不会触发页面的刷新，需要为其添加@State 装饰器，使其成为状态数据，它的改变会引起其所在组件的重新渲染，相关代码如下。

```
@Entry
@Component
struct FoodCategoryList {
  private foodItems: FoodData[] = initializeOnStartup()
  @State private showList: boolean = false
  build() {
    Stack({ alignContent: Alignment.TopEnd }) {
      if (this.showList) {
        FoodList({ foodItems: this.foodItems })
      } else {
```

```
            FoodCategory({ foodItems: this.foodItems })
        }
        Image( $r('app.media.Switch'))
            .height(24)
            .width(24)
            .margin({ top: 15, right: 10 })
            .onClick(() => {
                this.showList = !this.showList
            })
    }.height('100 % ')
  }
}
```

单击切换图标,FoodList 组件出现,再次单击,FoodList 组件消失。

(6) 创建显示所有食物的页签(All)。在 FoodCategory 组件内创建 Tabs 组件和其子组件 TabContent,设置 TabBar 为 All。设置 TabBars 的宽度为 280,布局模式为 Scrollable,即超过总长度后可以滑动。Tabs 是可以通过页签进行内容视图切换的容器组件,每个页签对应一个内容视图 TabContent,相关代码如下。

```
@Component
struct FoodCategory {
  private foodItems: FoodData[]
  build() {
    Stack() {
      Tabs() {
        TabContent() {}.tabBar('All')
      }
      .barWidth(280)
      .barMode(BarMode.Scrollable)
    }
  }
}
```

(7) 创建 FoodGrid 组件,作为 TabContent 的子组件,相关代码如下。

```
@Component
struct FoodGrid {
  private foodItems: FoodData[]
  build() {}
}
@Component
struct FoodCategory {
  private foodItems: FoodData[]
  build() {
    Stack() {
      Tabs() {
        TabContent() {
```

```
          FoodGrid({ foodItems: this.foodItems })
        }.tabBar('All')
      }
      .barWidth(280)
      .barMode(BarMode.Scrollable)
    }
  }
}
```

(8) 实现 2×6 的网格布局(一共 12 个食物数据资源)。创建 Grid 组件,设置列数 columnsTemplate('1fr 1fr'),行数 rowsTemplate('1fr 1fr 1fr 1fr 1fr 1fr'),行间距、列间距 rowsGap 和 columnsGap 均为 8。创建 Scroll 组件,使其可以滑动,相关代码如下。

```
@Component
struct FoodGrid {
  private foodItems: FoodData[]
  build() {
    Scroll() {
      Grid() {
        ForEach(this.foodItems, (item: FoodData) => {
          GridItem() {}
        }, (item: FoodData) => item.id.toString())
      }
      .rowsTemplate('1fr 1fr 1fr 1fr 1fr 1fr')
      .columnsTemplate('1fr 1fr')
      .columnsGap(8)
      .rowsGap(8)
    }
    .scrollBar(BarState.Off)
    .padding({left: 16, right: 16})
  }
}
```

(9) 创建 FoodGridItem 组件,展示食物图片、名称和卡路里,实现其 UI 布局,为 GridItem 的子组件。每个 FoodGridItem 高度为 184,行间距为 8,设置 Grid 总高度为(184+8)×6−8= 1144,相关代码如下。

```
@Component
struct FoodGridItem {
  private foodItem: FoodData
  build() {
    Column() {
      Row() {
        Image(this.foodItem.image)
          .objectFit(ImageFit.Contain)
          .height(152)
          .width('100%')
```

```
      }.backgroundColor('#FFf1f3f5')
      Flex({ justifyContent: FlexAlign.Start, alignItems: ItemAlign.Center }) {
        Text(this.foodItem.name)
          .fontSize(14)
          .flexGrow(1)
          .padding({ left: 8 })
        Text(this.foodItem.calories + 'kcal')
          .fontSize(14)
          .margin({ right: 6 })
      }
      .height(32)
      .width('100%')
      .backgroundColor('#FFe5e5e5')
    }
    .height(184)
    .width('100%')
  }
}
@Component
struct FoodGrid {
  private foodItems: FoodData[]
  build() {
    Scroll() {
      Grid() {
        ForEach(this.foodItems, (item: FoodData) => {
          GridItem() {
            FoodGridItem({foodItem: item})
          }
        }, (item: FoodData) => item.id.toString())
      }
      .rowsTemplate('1fr 1fr 1fr 1fr 1fr 1fr')
      .columnsTemplate('1fr 1fr')
      .columnsGap(8)
      .rowsGap(8)
      .height(1144)
    }
    .scrollBar(BarState.Off)
    .padding({ left: 16, right: 16 })
  }
}
```

(10) 创建展示蔬菜(Category.Vegetable)、水果(Category.Fruit)、坚果(Category.Nut)、海鲜(Category.SeaFood)和甜品(Category.Dessert)分类的页签,相关代码如下。

```
@Component
struct FoodCategory {
  private foodItems: FoodData[]
```

```
build() {
  Stack() {
    Tabs() {
      TabContent() {
        FoodGrid({ foodItems: this.foodItems })
      }.tabBar('All')
      TabContent() {
        FoodGrid({ foodItems: this.foodItems.filter(item => (item.category === Category.Vegetable)) })
      }.tabBar('Vegetable')
      TabContent() {
        FoodGrid({ foodItems: this.foodItems.filter(item => (item.category === Category.Fruit)) })
      }.tabBar('Fruit')
      TabContent() {
        FoodGrid({ foodItems: this.foodItems.filter(item => (item.category === Category.Nut)) })
      }.tabBar('Nut')
      TabContent() {
        FoodGrid({ foodItems: this.foodItems.filter(item => (item.category === Category.Seafood)) })
      }.tabBar('Seafood')
      TabContent() {
        FoodGrid({ foodItems: this.foodItems.filter(item => (item.category === Category.Dessert)) })
      }.tabBar('Dessert')
    }
    .barWidth(280)
    .barMode(BarMode.Scrollable)
  }
}
```

(11) 设置不同食物分类 Grid 的行数和高度。因为不同分类的食物数量不同,所以不能用'1fr 1fr 1fr 1fr 1fr 1fr '常量统一设置成 6 行。创建 gridRowTemplate 和 HeightValue,通过成员变量设置 Grid 行数和高度,相关代码如下。

```
@Component
struct FoodGrid {
  private foodItems: FoodData[]
  private gridRowTemplate : string = ''
  private heightValue: number
  build() {
    Scroll() {
      Grid() {
        ForEach(this.foodItems, (item: FoodData) => {
          GridItem() {
```

```
          FoodGridItem({foodItem: item})
        }
      }, (item: FoodData) => item.id.toString())
    }
    .rowsTemplate(this.gridRowTemplate)
    .columnsTemplate('1fr 1fr')
    .columnsGap(8)
    .rowsGap(8)
    .height(this.heightValue)
  }
  .scrollBar(BarState.Off)
  .padding({left: 16, right: 16})
}
```

调用 aboutToAppear 接口计算行数(gridRowTemplate)和高度(heightValue)。

```
aboutToAppear() {
  var rows = Math.round(this.foodItems.length / 2);
  this.gridRowTemplate = '1fr '.repeat(rows);
  this.heightValue = rows * 192 - 8;
}
```

自定义组件提供两个生命周期的回调接口 aboutToAppear 和 aboutToDisappear。aboutToAppear 的执行时机在创建自定义组件后、执行 Build 方法之前。aboutToDisappear 在自定义组件去初始化的时机执行。自定义组件的回调接口流程如图 6-6 所示。

相关代码如下。

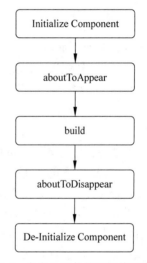

图 6-6 自定义组件的回调接口流程

```
@Component
struct FoodGrid {
  private foodItems: FoodData[]
  private gridRowTemplate : string = ''
  private heightValue: number
  aboutToAppear() {
    var rows = Math.round(this.foodItems.length / 2);
    this.gridRowTemplate = '1fr '.repeat(rows);
    this.heightValue = rows * 192 - 8;
  }
  build() {
    Scroll() {
      Grid() {
        ForEach(this.foodItems, (item: FoodData) => {
          GridItem() {
```

```
          FoodGridItem({foodItem: item})
        }
      }, (item: FoodData) => item.id.toString())
    }
    .rowsTemplate(this.gridRowTemplate)
    .columnsTemplate('1fr 1fr')
    .columnsGap(8)
    .rowsGap(8)
    .height(this.heightValue)
  }
  .scrollBar(BarState.Off)
  .padding({left: 16, right: 16})
 }
}
```

6.3.4 页面跳转与数据传递

本节将介绍页面跳转和数据传递的两个功能。页面跳转：单击食物分类列表页面的食物条目后，跳转到食物详情页，单击食物详情页的返回按钮，返回到食物列表页。页面间数据传递：单击不同的食物条目后，FoodDetail 接收前一个页面的数据，渲染对应的食物详情页。

1. 页面跳转

声明式 UI 范式提供了两种机制实现页面间的跳转：路由容器组件 Navigator，包装了页面路由的能力，指定页面 target 后，使其包裹的子组件都具有路由能力；路由 RouterAPI 接口在页面上引入 router，可以调用 router 的各种接口，从而实现页面路由的各种操作。下面分别学习两种跳转机制实现食物分类列表页面和食物详情页的链接。

(1) 单击 FoodListItem 后跳转到 FoodDetail 页面。在 FoodListItem 内创建 Navigator 组件，使其子组件都具有路由功能，目标页面 target 为"pages/FoodDetail"，相关代码如下。

```
@Component
struct FoodListItem {
  private foodItem: FoodData
  build() {
    Navigator({ target: 'pages/FoodDetail' }) {
      Flex({ justifyContent: FlexAlign.Start, alignItems: ItemAlign.Center }) {
        Image(this.foodItem.image)
          .objectFit(ImageFit.Contain)
          .height(40)
          .width(40)
          .backgroundColor('#FFf1f3f5')
          .margin({ right: 16 })
        Text(this.foodItem.name)
          .fontSize(14)
          .flexGrow(1)
```

```
        Text(this.foodItem.calories + 'kcal')
          .fontSize(14)
      }
      .height(64)
    }
    .margin({ right: 24, left:32 })
  }
}
```

(2) 单击 FoodGridItem 后跳转到 FoodDetail 页面。调用页面路由 router 模块的 push 接口，将 FoodDetail 页面推送到路由栈中，实现页面跳转。使用 router 路由 API 接口，需要先引入 router，相关代码如下。

```
import router from '@system.router'
@Component
struct FoodGridItem {
  private foodItem: FoodData
  build() {
    Column() {
      ......
    }
    .height(184)
    .width('100%')
    .onClick(() => {
      router.push({ uri: 'pages/FoodDetail' })
    })
  }
}
```

(3) 在 FoodDetail 页面增加回到食物列表页面的图标。在 resources→phone→media 文件夹下存入后退图标 Back.png。新建自定义组件 PageTitle，包含后退的图标和 Food Detail 的文本，调用路由的 router.back() 接口，弹出路由栈最上面的页面，即返回上一级页面，相关代码如下。

```
//FoodDetail.ets
import router from '@system.router'
@Component
struct PageTitle {
    build() {
        Flex({ alignItems: ItemAlign.Start }) {
            Image( $r('app.media.Back'))
                .width(21.8)
                .height(19.6)
            Text('Food Detail')
                .fontSize(21.8)
                .margin({left: 17.4})
```

```
            }
            .height(61)
            .backgroundColor('#FFedf2f5')
            .padding({ top: 13, bottom: 15, left: 28.3 })
            .onClick(() => {
                router.back()
            })
        }
    }
```

(4) 在 FoodDetail 组件内创建 Stack 组件,包含 FoodImageDisplay 和 PageTitle 子组件,设置其对齐方式为左上对齐 TopStart,相关代码如下。

```
@Entry
@Component
struct FoodDetail {
    build() {
        Column() {
            Stack( { alignContent: Alignment.TopStart }) {
                FoodImageDisplay()
                PageTitle()
            }
            ContentTable()
        }
        .alignItems(HorizontalAlign.Center)
    }
}
```

2. 页面间数据传递

完成 FoodCategoryList 和 FoodDetail 页面的跳转、回退后,单击不同的 FoodListItem/FoodGridItem,跳转的 FoodDetail 页面都是西红柿 Tomato 的详细介绍,这是因为没有构建起两个页面的数据传递,需要用到携带参数(parameter)路由。

(1) 在 FoodListItem 组件的 Navigator 设置其 params 属性,params 属性接收 key-value 的 Object,相关代码如下。

```
//FoodList.ets
@Component
struct FoodListItem {
    private foodItem: FoodData
    build() {
        Navigator({ target: 'pages/FoodDetail' }) {
            ......
        }
        .params({ foodData: this.foodItem })
    }
}
```

FoodGridItem 调用的 routerAPI 同样有携带参数跳转的能力,使用方法和 Navigator 类似。

```
router.push({
  uri: 'pages/FoodDetail',
  params: { foodData: this.foodItem }
})
```

(2) FoodDetail 页面引入 FoodData 类,在 FoodDetail 组件内添加 foodItem 成员变量,相关代码如下。

```
//FoodDetail.ets
import { FoodData } from '../model/FoodData'
@Entry
@Component
struct FoodDetail {
  private foodItem: FoodData
  build() {
    ......
  }
}
```

(3) 获取 FoodData 对应的 Value。调用 router.getParams().foodData 获取 FoodCategoryList 页面跳转来时携带的 FoodDate 对应数据,相关代码如下。

```
@Entry
@Component
struct FoodDetail {
  private foodItem: FoodData = router.getParams().foodData
  build() {
    ......
  }
}
```

(4) 重构 FoodDetail 页面的组件。在构建视图时,FoodDetail 页面的食物信息是直接声明的常量,需要传递 FoodData 数据对其进行重新赋值,相关代码如下。

```
@Component
struct PageTitle {
  build() {
    Flex({ alignItems: ItemAlign.Start }) {
      Image( $r('app.media.Back'))
        .width(21.8)
        .height(19.6)
      Text('Food Detail')
        .fontSize(21.8)
        .margin({left: 17.4})
```

```
            }
            .height(61)
            .backgroundColor('#FFedf2f5')
            .padding({ top: 13, bottom: 15, left: 28.3 })
            .onClick(() => {
                router.back()
            })
        }
    }
}
@Component
struct FoodImageDisplay {
    private foodItem: FoodData
    build() {
        Stack({ alignContent: Alignment.BottomStart }) {
            Image(this.foodItem.image)
                .objectFit(ImageFit.Contain)
            Text(this.foodItem.name)
                .fontSize(26)
                .fontWeight(500)
                .margin({ left: 26, bottom: 17.4 })
        }
        .height(357)
        .backgroundColor('#FFedf2f5')
    }
}
@Component
struct ContentTable {
    private foodItem: FoodData
    @Builder IngredientItem(title:string, name: string, value: string) {
        Flex() {
            Text(title)
                .fontSize(17.4)
                .fontWeight(FontWeight.Bold)
                .layoutWeight(1)
            Flex() {
                Text(name)
                    .fontSize(17.4)
                    .flexGrow(1)
                Text(value)
                    .fontSize(17.4)
            }
            .layoutWeight(2)
        }
    }
    build() {
        Flex ({ direction: FlexDirection.Column, justifyContent: FlexAlign.SpaceBetween,
alignItems: ItemAlign.Start }) {
```

```
        this.IngredientItem('Calories', 'Calories', this.foodItem.calories + 'kcal')
        this.IngredientItem('Nutrition', 'Protein', this.foodItem.protein + 'g')
        this.IngredientItem('', 'Fat', this.foodItem.fat + 'g')
        this.IngredientItem('', 'Carbohydrates', this.foodItem.carbohydrates + 'g')
        this.IngredientItem('', 'VitaminC', this.foodItem.vitaminC + 'mg')
      }
      .height(280)
      .padding({ top: 30, right: 30, left: 30 })
  }
}
@Entry
@Component
struct FoodDetail {
  private foodItem: FoodData = router.getParams().foodData
  build() {
    Column() {
      Stack( { alignContent: Alignment.TopStart }) {
        FoodImageDisplay({ foodItem: this.foodItem })
        PageTitle()
      }
      ContentTable({ foodItem: this.foodItem })
    }
    .alignItems(HorizontalAlign.Center)
  }
}
```

6.4 绘图和动画

绘图和动画是应用开发中常用的技术手段,本节介绍其开发方法。参考地址为: https://gitee.com/harmonyos/harmonyos_app_samples/tree/master/ETSUI/eTSDrawingAndAnimation,示例构建健康饮食应用的闪屏动画,展示绘制组件和显式动画的基本用法。

6.4.1 绘制图形

绘制能力主要通过绘制组件支撑,支持 svg 标准绘制命令。本节主要学习如何使用绘制组件、绘制详情页食物成分标签(基本几何图形)和应用 Logo(自定义图形)。

1. 绘制基本几何图形

绘制组件封装了一些常见的基本几何图形,例如矩形(Rect)、圆形(Circle)、椭圆形(Ellipse)等,为开发者省去了路线计算的过程。FoodDetail 页面的食物成分表,给每项成分名称前都加上一个圆形的图标作为成分标签。

(1) 创建 Circle 组件,在每项含量成分前增加一个圆形图标作为标签,设置 Circle 的直径为 6vp。修改 FoodDetail 页面的 ContentTable 组件里的 IngredientItem 方法,在成分名称前添加 Circle,相关代码如下。

```
//FoodDetail.ets
@Component
struct ContentTable {
  private foodItem: FoodData
  @Builder IngredientItem(title:string, colorValue: string, name: string, value: string) {
    Flex() {
      Text(title)
        .fontSize(17.4)
        .fontWeight(FontWeight.Bold)
        .layoutWeight(1)
      Flex({ alignItems: ItemAlign.Center }) {
        Circle({width: 6, height: 6})
          .margin({right: 12})
          .fill(colorValue)
        Text(name)
          .fontSize(17.4)
          .flexGrow(1)
        Text(value)
          .fontSize(17.4)
      }
      .layoutWeight(2)
    }
  }
  build() {
    ......
  }
}
```

(2) 每个成分的标签颜色不同,所以在 Build 方法中,调用 IngredientItem,给每个 Circle 填充不同的颜色,相关代码如下。

```
//FoodDetail.ets
@Component
struct ContentTable {
  private foodItem: FoodData
  @Builder IngredientItem(title:string, colorValue: string, name: string, value: string) {
    Flex() {
      Text(title)
        .fontSize(17.4)
        .fontWeight(FontWeight.Bold)
        .layoutWeight(1)
      Flex({ alignItems: ItemAlign.Center }) {
        Circle({width: 6, height: 6})
          .margin({right: 12})
```

```
                .fill(colorValue)
            Text(name)
                .fontSize(17.4)
                .flexGrow(1)
            Text(value)
                .fontSize(17.4)
        }
        .layoutWeight(2)
    }
}
build() {
    Flex ({ direction: FlexDirection.Column, justifyContent: FlexAlign.SpaceBetween,
alignItems: ItemAlign.Start }) {
        this.IngredientItem('Calories', '#FFf54040', 'Calories', this.foodItem.calories + 'kcal')
        this.IngredientItem('Nutrition', '#FFcccccc', 'Protein', this.foodItem.protein + 'g')
        this.IngredientItem('', '#FFf5d640', 'Fat', this.foodItem.fat + 'g')
        this.IngredientItem('', '#FF9e9eff', 'Carbohydrates', this.foodItem.carbohydrates + 'g')
        this.IngredientItem('', '#FF53f540', 'VitaminC', this.foodItem.vitaminC + 'mg')
    }
    .height(280)
    .padding({ top: 30, right: 30, left: 30 })
}
}
```

2. 绘制自定义几何图形

除绘制基础几何图形，还可以使用 Path 组件绘制自定义的路线，下面绘制应用的 Logo 图案。

(1) 在 pages 文件夹下创建新的 eTS 页面 Logo.ets，对应在 config.json 中的 pages 标签下添加 Logo.ets，相关代码如下。

```
"pages": [
  "pages/Logo",
  "pages/FoodCategoryList",
  "pages/FoodDetail"
],
```

(2) 删除 Logo.ets 中的模板代码，创建 Logo Component，相关代码如下。

```
@Entry
@Component
struct Logo {
  build() {
  }
}
```

(3) 创建 Flex 组件为根节点，宽高设置为 100%，设置其在主轴方向和交叉轴方向的对齐方式都为 Center，创建 Shape 组件为 Flex 子组件。

Shape 组件是所有绘制组件的父组件。如果将组合多个绘制组件成为一个整体，需要创建 Shape 作为其父组件。

要绘制的 Logo 大小为 630px * 630px。声明式 UI 范式支持多种长度单位的设置，可以直接使用 number 作为参数，采用默认长度单位 vp。vp 和设备分辨率及屏幕密度有关。例如，设备分辨率为 1176 * 2400，屏幕基准密度（resolution）为 3，vp＝px / resolution，则该设备屏幕宽度是 392vp。

绘制组件采用 svg 标准，默认采取 px 为单位，为方便统一，在绘制 Logo 部分，采取 px 为单位。声明式 UI 框架同样也支持 px 单位，入参类型 String，设置宽度为 630px，即 210vp，设置方式为 width('630px')或者 width(210)，相关代码如下。

```
@Entry
@Component
struct Logo {
  build() {
    Flex({ alignItems: ItemAlign.Center, justifyContent: FlexAlign.Center }) {
      Shape() {
      }
      .height('630px')
      .width('630px')
    }
    .width('100%')
    .height('100%')
  }
}
```

（4）给页面填充渐变色。设置为线性渐变，偏移角度为 180deg，三段渐变 ♯BDE895→95DE7F→♯7AB967，其区间分别为[0,0.1]，[0.1,0.6]，[0.6,1]，相关代码如下。

```
.linearGradient(
  {
    angle: 180,
    colors: [['#BDE895', 0.1], ["#95DE7F", 0.6], ["#7AB967", 1]]
  })
@Entry
@Component
struct Logo {
  build() {
    Flex({ alignItems: ItemAlign.Center, justifyContent: FlexAlign.Center }) {
      Shape() {
      }
      .height('630px')
      .width('630px')
    }
    .width('100%')
    .height('100%')
```

```
        .linearGradient(
        {
          angle: 180,
          colors: [['#BDE895', 0.1], ["#95DE7F", 0.6], ["#7AB967", 1]]
        })
      }
    }
```

(5) 绘制第一条路线 Path，设置其绘制命令。

```
Path()
  .commands('M162 128.7 a222 222 0 0 1 100.8 374.4 H198 a36 36 0 0 3 -36 -36 V128.7 z')
```

Path 的绘制命令采用 svg 标准，上述命令可分解如下。

(1) M162 128.7：将笔触移动到(Moveto)坐标点(162,128.7)。

(2) a222 222 0 0 1 100.8 374.4：画圆弧线(elliptical arc)半径 rx、ry 均为 222，x 轴旋转角度 x-axis-rotation 为 0，角度大小 large-arc-flag 为 0，即小弧度角，弧线方向(sweep-flag)为 1，即逆时针画弧线，小写 a 为相对位置，即终点坐标为(162+100.8=262.8,128.7+374.4=503.1)。

(3) H198：画水平线(horizontal lineto)到 198，画(262.8,503.1)到(198,503.1)的水平线。

(4) a36 36 0 0 3 -36 -36：画圆弧线(elliptical arc)，含义同上，结束点为(198-36=162,503.1-36=467.1)。

(5) V128.7：画垂直线(vertical lineto)到 128.7，即画(162,467.1)到(162,128.7)的垂直线。

(6) z：关闭路径(closepath)。

(7) 填充颜色为白色，相关代码如下。

```
.fill(Color.White)
@Entry
@Component
struct Logo {
  build() {
    Flex({ alignItems: ItemAlign.Center, justifyContent: FlexAlign.Center }) {
      Shape() {
        Path()
          .commands('M162 128.7 a222 222 0 0 1 100.8 374.4 H198 a36 36 0 0 3 -36 -36')
          .fill(Color.White)
      }
      .height('630px')
      .width('630px')
    }
    .width('100%')
    .height('100%')
```

```
    .linearGradient(
      {
        angle: 180,
        colors: [['#BDE895', 0.1], ["#95DE7F", 0.6], ["#7AB967", 1]]
      })
  }
}
```

(8) 在 Shape 组件内绘制第二个 Path。第二个 Path 的背景色为渐变色,但是渐变色的填充是其整体的 box,所以需要 clip 将其裁剪,入参为 Shape,即按照 Shape 的形状进行裁剪,相关代码如下。

```
Path()
  .commands('M319.5 128.1 c103.5 0 187.5 84 187.5 187.5 v15 a172.5 172.5 0 0 3 -172.5 172.5 H198 a36 36 0 0 3 -13.8 -1 207 207 0 0 0 87 -372 h48.3 z')
  .fill('none')
  .linearGradient(
  {
    angle: 30,
    colors: [["#C4FFA0", 0], ["#ffffff", 1]]
  })
  .clip(new Path().commands('M319.5 128.1 c103.5 0 187.5 84 187.5 187.5 v15 a172.5 172.5 0 0 3 -172.5 172.5 H198 a36 36 0 0 3 -13.8 -1 207 207 0 0 0 87 -372 h48.3 z'))
```

Path 的绘制命令比较长,可以将其作为组件的成员变量,通过 this 调用。

```
@Entry
@Component
struct Logo {
  private pathCommands1:string = 'M319.5 128.1 c103.5 0 187.5 84 187.5 187.5 v15 a172.5 172.5 0 0 3 -172.5 172.5 H198 a36 36 0 0 3 -13.8 -1 207 207 0 0 0 87 -372 h48.3 z'
  build() {
    ……
      Path()
        .commands(this.pathCommands1)
        .fill('none')
        .linearGradient(
        {
          angle: 30,
          colors: [["#C4FFA0", 0], ["#ffffff", 1]]
        })
        .clip(new Path().commands(this.pathCommands1))
      ……
  }
}
```

(9) 在 Shape 组件内绘制第二个 Path，相关代码如下。

```
@Entry
@Component
struct Logo {
  private pathCommands1:string = 'M319.5 128.1 c103.5 0 187.5 84 187.5 187.5 v15 a172.5 172.5 0 0 3 -172.5 172.5 H198 a36 36 0 0 3 -13.8 -1 207 207 0 0 0 87 -372 h48.3 z'
  private pathCommands2:string = 'M270.6 128.1 h48.6 c51.6 0 98.4 21 132.3 54.6 a411 411 0 0 3 -45.6 123 c-25.2 45.6 -56.4 84 -87.6 110.4 a206.1 206.1 0 0 0 -47.7 -288 z'
  build() {
    Flex({ alignItems: ItemAlign.Center, justifyContent: FlexAlign.Center }) {
      Shape() {
        Path()
          .commands('M162 128.7 a222 222 0 0 1 100.8 374.4 H198 a36 36 0 0 3 -36 -36')
          .fill(Color.White)
        Path()
          .commands(this.pathCommands1)
          .fill('none')
          .linearGradient(
          {
            angle: 30,
            colors: [["#C4FFA0", 0], ["#ffffff", 1]]
          })
          .clip(new Path().commands(this.pathCommands1))
        Path()
          .commands(this.pathCommands2)
          .fill('none')
          .linearGradient(
          {
            angle: 50,
            colors: [['#8CC36A', 0.1], ["#B3EB90", 0.4], ["#ffffff", 0.7]]
          })
          .clip(new Path().commands(this.pathCommands2))
      }
      .height('630px')
      .width('630px')
    }
    .width('100%')
    .height('100%')
    .linearGradient(
    {
      angle: 180,
      colors: [['#BDE895', 0.1], ["#95DE7F", 0.6], ["#7AB967", 1]]
    })
  }
}
```

完成应用 Logo 的绘制。Shape 组合了三个 Path 组件,通过 svg 命令绘制出一片叶子,寓意绿色健康饮食方式。

(10) 添加应用标题和标语,相关代码如下。

```
@Entry
@Component
struct Logo {
  private pathCommands1:string = 'M319.5 128.1 c103.5 0 187.5 84 187.5 187.5 v15 a172.5 172.5 0 0 3 -172.5 172.5 H198 a36 36 0 0 3 -13.8 -1 207 207 0 0 0 87 -372 h48.3 z'
  private pathCommands2:string = 'M270.6 128.1 h48.6 c51.6 0 98.4 21 132.3 54.6 a411 411 0 0 3 -45.6 123 c-25.2 45.6 -56.4 84 -87.6 110.4 a206.1 206.1 0 0 0 -47.7 -288 z'
  build() {
    Flex({ direction: FlexDirection.Column, alignItems: ItemAlign.Center, justifyContent: FlexAlign.Center }) {
      Shape() {
        Path()
          .commands('M162 128.7 a222 222 0 0 1 100.8 374.4 H198 a36 36 0 0 3 -36 -36')
          .fill(Color.White)
        Path()
          .commands(this.pathCommands1)
          .fill('none')
          .linearGradient(
          {
            angle: 30,
            colors: [["#C4FFA0", 0], ["#ffffff", 1]]
          })
          .clip(new Path().commands(this.pathCommands1))
        Path()
          .commands(this.pathCommands2)
          .fill('none')
          .linearGradient(
          {
            angle: 50,
            colors: [['#8CC36A', 0.1], ["#B3EB90", 0.4], ["#ffffff", 0.7]]
          })
          .clip(new Path().commands(this.pathCommands2))
      }
      .height('630px')
      .width('630px')
      Text('Healthy Diet')
        .fontSize(26)
        .fontColor(Color.White)
        .margin({ top:300 })
      Text('Healthy life comes from a balanced diet')
        .fontSize(17)
        .fontColor(Color.White)
```

```
        .margin({ top:4 })
    }
    .width('100%')
    .height('100%')
    .linearGradient(
      {
        angle: 180,
        colors: [['#BDE895', 0.1], ["#95DE7F", 0.6], ["#7AB967", 1]]
      })
  }
}
```

绘制图形后的页面效果如图 6-7 所示。

图 6-7　页面绘制效果

6.4.2　添加动画效果

动画功能主要包含组件动画和页面间动画,并开放了插值计算和矩阵变换的动画能力接口,开发者可以自主设计。

本节主要完成两个动画效果:启动页的闪屏动画,即 Logo 图标的渐出和放大效果;食物列表页和食物详情页的共享元素转场动画效果。

1. animateTo 实现闪屏动画

声明式 UI 范式组件动画包括属性动画和 animateTo 显式动画。属性动画:设置组件通用属性变化的动画效果。显式动画:可以设置组件从状态 A 到状态 B 的变化动画效果,包括样式、位置信息和节点的增加删除等,开发者无须关注变化过程,只需指定起点和终点的状态。animateTo 还提供播放状态的回调接口,是对属性动画的增强与封装。

闪屏页面的动画效果是 Logo 图标的渐出和放大效果的动画,动画结束后跳转到食物分类列表页面。下面通过 animateTo 实现启动页动画的闪屏效果。

(1) 动画效果自动播放。闪屏动画的预期效果是进入 Logo 页面后, animateTo 动画效果自动开始播放, 可以借助组件显隐事件的回调接口实现。调用 Shape 的 onAppear 方法, 设置其显式动画, 相关代码如下。

```
Shape() {
  ...
}
.onAppear(() => {
  animateTo()
})
```

(2) 创建 opacity 和 scale 数值的成员变量, 用装饰器 @State 修饰。表示其为有状态的数据, 即改变会触发页面的刷新, 相关代码如下。

```
@Entry
@Component
struct Logo {
  @State private opacityValue: number = 0
  @State private scaleValue: number = 0
  build() {
    Shape() {
      ...
    }
    .scale({ x: this.scaleValue, y: this.scaleValue })
    .opacity(this.opacityValue)
    .onAppear(() => {
      animateTo()
    })
  }
}
```

(3) 设置 animateTo 的动画曲线 curve。Logo 的加速曲线为先慢后快, 使用贝塞尔曲线 cubicBezier, cubicBezier(0.4, 0, 1, 1)。需要通过动画能力接口中的插值计算, 导入 curves 模块。

```
import Curves from '@ohos.curves'
```

@ohos.curves 模块提供了线性 Curve、Linear、阶梯 step、三阶贝塞尔 (cubicBezier) 和弹簧 (spring) 插值曲线的初始化函数, 可以根据入参创建一个插值曲线对象, 相关代码如下。

```
@Entry
@Component
struct Logo {
  @State private opacityValue: number = 0
  @State private scaleValue: number = 0
```

```
    private curve1 = Curves.cubicBezier(0.4, 0, 1, 1)
    build() {
      Shape() {
        ...
      }
      .scale({ x: this.scaleValue, y: this.scaleValue })
      .opacity(this.opacityValue)
      .onAppear(() => {
        animateTo({
            curve: this.curve1
        })
      })
    }
  }
```

(4) 设置动画时长为 1s，延时 0.1s 开始播放，设置显示动效 event 的闭包函数，即起点状态到终点状态为透明度 opacityValue 和大小 scaleValue 从 0～1 变化，实现 Logo 的渐出和放大效果，相关代码如下。

```
@Entry
@Component
struct Logo {
  @State private opacityValue: number = 0
  @State private scaleValue: number = 0
  private curve1 = Curves.cubicBezier(0.4, 0, 1, 1)
  build() {
    Shape() {
      ...
    }
    .scale({ x: this.scaleValue, y: this.scaleValue })
    .opacity(this.opacityValue)
    .onAppear(() => {
      animateTo({
        duration: 2000,
        curve: this.curve1,
        delay: 100,
      }, () => {
        this.opacityValue = 1
        this.scaleValue = 1
      })
    })
  }
}
```

(5) 闪屏动画播放结束后定格 1s，进入 FoodCategoryList 页面。设置 animateTo 的 onFinish 回调接口，调用定时器 Timer 的 setTimeout 接口延时 1s 后，调用 router.replace，

显示 FoodCategoryList 页面,相关代码请扫描二维码获取。

2. 页面转场动画

食物分类列表页和食物详情页之间的共享元素转场,即单击 FoodListItem/FoodGridItem 后,食物缩略图会放大,随着页面跳转到食物详情页的大图。

(1)设置 FoodListItem 和 FoodGridItem 的 Image 组件的共享元素转场方法 (sharedTransition)。转场 ID 为 foodItem.id,转场动画时长为 1s,延时 0.1s 播放,变化曲线为贝塞尔曲线 Curves.cubicBezier(0.2,0.2,0.1,1.0),需引入 curves 模块。

共享转场时会携带当前元素被设置的属性,所以创建 Row 组件,使其作为 Image 的父组件,设置背景颜色在 Row 上。

在 FoodListItem 的 Image 组件上,为了转场动画的流畅,设置 autoResize 为 false,因为 image 组件默认会根据最终展示的区域调整图源的大小,以优化图片渲染性能。在转场动画时,图片在放大的过程中会被重新加载,相关代码如下。

```
//FoodList.ets
import Curves from '@ohos.curves'
@Component
struct FoodListItem {
  private foodItem: FoodData
  build() {
    Navigator({ target: 'pages/FoodDetail' }) {
      Flex({ justifyContent: FlexAlign.Start, alignItems: ItemAlign.Center }) {
        Row() {
          Image(this.foodItem.image)
            .objectFit(ImageFit.Contain)
            .autoResize(false)
            .height(40)
            .width(40)
            .sharedTransition(this.foodItem.id, { duration: 1000, curve: Curves.cubicBezier(0.2, 0.2, 0.1, 1.0), delay: 100 })
        }
        .backgroundColor('#FFf1f3f5')
        .margin({ right: 16 })
        Text(this.foodItem.name)
          .fontSize(14)
          .flexGrow(1)
        Text(this.foodItem.calories + 'kcal')
          .fontSize(14)
      }
      .height(64)
    }
    .params({ foodData: this.foodItem })
    .margin({ right: 24, left:32 })
  }
```

```
}
@Component
struct FoodGridItem {
  private foodItem: FoodData
  build() {
    Column() {
      Row() {
        Image(this.foodItem.image)
          .objectFit(ImageFit.Contain)
          .height(152)
          .width('100%')
          .sharedTransition(this.foodItem.id, { duration: 1000, curve: Curves.cubicBezier
(0.2, 0.2, 0.1, 1.0), delay: 100 })
      }.backgroundColor('#FFf1f3f5')
      Flex({ justifyContent: FlexAlign.Start, alignItems: ItemAlign.Center }) {
        Text(this.foodItem.name)
          .fontSize(14)
          .flexGrow(1)
          .padding({ left: 8 })
        Text(this.foodItem.calories + 'kcal')
          .fontSize(14)
          .margin({ right: 6 })
      }
      .height(32)
      .width('100%')
      .backgroundColor('#FFe5e5e5')
    }
    .height(184)
    .width('100%')
    .onClick(() => {
      router.push({ uri: 'pages/FoodDetail', params: { foodId: this.foodItem } })
    })
  }
}
```

(2) 设置 FoodDetail 页面 FoodImageDisplay 的 Image 组件的共享元素转场方法 (sharedTransition)，设置方法同上，相关代码如下。

```
import Curves from '@ohos.curves'
@Component
struct FoodImageDisplay {
  private foodItem: FoodData
  build() {
    Stack({ alignContent: Alignment.BottomStart }) {
      Image(this.foodItem.image)
        .objectFit(ImageFit.Contain)
        .sharedTransition(this.foodItem.id, { duration: 1000, curve: Curves.cubicBezier
```

```
        (0.2, 0.2, 0.1, 1.0), delay: 100 })
        Text(this.foodItem.name)
          .fontSize(26)
          .fontWeight(500)
          .margin({ left: 26, bottom: 17.4 })
      }
      .height(357)
      .backgroundColor('#FFedf2f5')
    }
  }
```

声明式 UI 框架提供了丰富的动效接口,合理地应用和组合可以使应用更具有设计感。

第 7 章 贪吃蛇小游戏

本项目通过鸿蒙系统开发工具 DevEco Studio，使用 JavaScript 软件，开发一款贪吃蛇游戏，实现积分系统。

7.1 总体设计

本部分包括系统架构和系统流程。

7.1.1 系统架构

系统架构如图 7-1 所示。

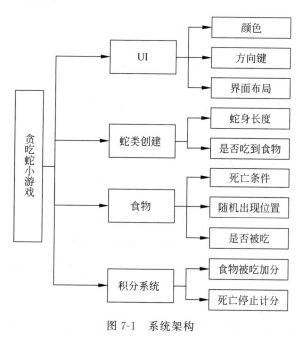

图 7-1　系统架构

7.1.2　系统流程

系统流程如图 7-2 所示。

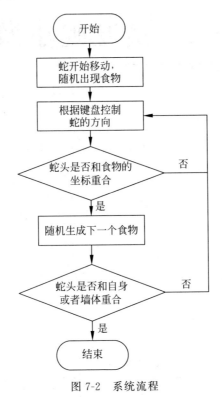

图 7-2　系统流程

7.2　开发环境

本部分包括开发工具及开发语言。

7.2.1　开发工具

本实验使用 DevEco Studio 开发工具,安装过程如下。

(1) 注册华为开发者账号,按照指引完成注册并登录,在官网下载 DevEco Studio 并安装,参考教程地址为:https://developer.harmonyos.com/cn/docs/documentation/doc-guides/tools_overview-0000001053582387。

(2) 下载并安装 Node.js。

(3) 新建项目,设备类型和模板在 device 中选择 Phone,然后选择 Empty Feature Ability(JS),单击 Next,并填写相关信息。

（4）创建后的应用目录结构如图 7-3 所示，与 Android Studio 结构相似。

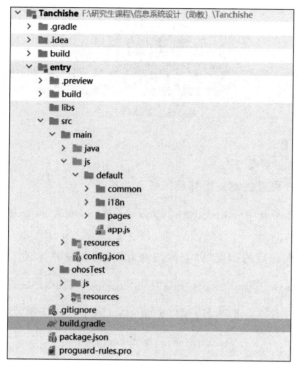

图 7-3　应用目录结构

（5）在 src/main/js 目录下进行贪吃蛇小游戏的开发。

7.2.2　开发语言

DevEco Studio 支持多种语言的代码开发和调试，包括 Java、XML（Extensible Markup Language）、C/C++、JS（JavaScript）、CSS（Cascading Style Sheets）和 HML（HarmonyOS Markup Language）。本项目使用 JS 进行开发，需要提前安装 Node.js。

7.3　开发实现

本项目包括 UI 设计开发和逻辑代码程序开发，下面分别给出各模块的功能介绍及相关代码。

7.3.1　UI 设计开发

本部分包括图片导入、UI 开发和完整 UI 代码。

1. 图片导入

将选好的界面图片导入 project 中，然后将选好作为控制蛇头四个方向的图片文件

(.png 格式)放入 js/default/common 文件夹下,如图 7-4 所示。

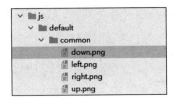

图 7-4 导入图片

2. UI 开发

贪吃蛇游戏的 UI 设计如下。

(1) 使用画布组件设置贪吃蛇的移动区域。

```
< canvas ref = "canvasref" style = "width: 350px; height: 400px; background - color: black;">
</canvas >
```

(2) 设置四个方向按键的布局(以上面按键为例,其余按键类似)。

```
< image src = "/common/up.png"class = "backBtnup"onclick = "onStartGame(1)"></image >
```

(3) 使用 CSS 选择器,实现对 HTML 页面元素(四个方向按键元素)的控制。

```
.backBtnup, .backBtncenter, .backBtnleft, .backBtnright {
    width: 40px;
    height: 40px;
    margin - bottom: 10px;
    margin - top: 0px;
    border - radius: 10px;
    background - color: black;
}
```

(4) 判断游戏是否结束,若结束显示 gameOver,若未结束,则记录分数。

```
< text if = "{{gameOver}}" class = "scoretitle">
        < span > Game Over!!!</span >
    </text >
< text if = "{{!gameOver}}" class = "scoretitle">
        < span > Score: {{score}}</span >
    </text >
```

3. 完整 UI 代码

完整 UI 设计代码如下。

1) HML 文件

```
<!-- 容器 -->
< div class = "container">
<!-- 标题 -->
```

```html
        <text class = "title"> Snake Game </text>
<!-- 画布组件:贪吃蛇的移动区域 -->
        <canvas ref = "canvasref" style = "width: 350px; height: 400px; background-color: black;"></canvas>
<!-- 上按键 -->
        <image src = "/common/up.png" class = "backBtnup" onclick = "onStartGame(1)"></image>
<!-- 下面三个按键用同一样式,所以用同一个div包围 -->
        <div class = "directsecond">
            <!-- 左按键 -->
            <image src = "/common/left.png" class = "backBtnleft" onclick = "onStartGame(2)"></image>
            <!-- 下按键 -->
            <image src = "/common/down.png" class = "backBtncenter" onclick = "onStartGame(3)"></image>
            <!-- 右按键 -->
            <image src = "/common/right.png" class = "backBtnright" onclick = "onStartGame(4)"></image>
        </div>
<!-- 用if判断,如果游戏结束,则显示该模块 -->
        <text if = "{{gameOver}}" class = "scoretitle">
            <span> Game Over!!!
</span>
        </text>
<!-- 用if判断,如果游戏没有结束,则显示该模块.显示得分 -->
        <text if = "{{!gameOver}}" class = "scoretitle">
            <span> Score: {{score}}</span>
        </text>
</div>
```

2) CSS 文件

```css
.container {
    flex-direction: column;
    justify-content: center;
    align-items: center;
    background-color: white;
}
/*
.title {
    font-size: 10px;
    margin-bottom: 10px;
}
*/
.scoretitle {
    font-size: 30px;
    margin-top: 20px;
}

/*
    css选择器,逗号代表并列关系
```

```css
*/
.backBtnup, .backBtncenter, .backBtnleft, .backBtnright {
    width: 40px;
    height: 40px;
    margin-bottom: 10px;
    margin-top: 0px;
    border-radius: 10px;
    background-color: black;
}
.backBtnup {
    margin-top: 10px;
}
.backBtncenter {
    margin-left: 10px;
    margin-right: 10px;
}
.directsecond {
    flex-direction: row;
    justify-content: center;
    align-items: center;
}
```

7.3.2 程序代码开发

本部分包括程序初始化、创建贪吃蛇体、随机食物位置、碰壁、吃食物及完整逻辑代码。

1. 程序初始化

对游戏背景宽度、颜色、按键状态、游戏状态、食物位置等多个数据进行初始化设置,相关代码如下。

```
data: {
    title: "",
    snakeSize: 10,      //蛇身格子像素大小
    w: 350,             //背景宽度
    h: 400,             //背景高度
    score: 0,           //得分为0
    snake : [],         //数组用来存放每个格子的位置
    ctx: null,          //调用填充颜色
    food: null,         //食物位置
    direction: '',      //按键状态
    gameOver: false,    //游戏状态
    tail: {             //记录更新后蛇头的位置
        x: 0,
        y: 0
    }
,
    interval : null     //获得setInterval()的返回值
},
```

2．创建贪吃蛇体

默认初始蛇的长度为7，初始化在左上角，将x轴和y轴的坐标数据存储到数组中，即每条蛇占用格子的位置，相关代码如下。

```
drawSnake() {
    var len = 7;
    var snake = [];
    //默认蛇的长度为7
    for (var i = len - 1; i >= 0; i--) {
        //将x轴和y轴的坐标数据存到数组中,这些数据是蛇所在格子的位置
        snake.push({
            x: 0,
            y: i
        });
    }
    //更新蛇的长度
    this.snake = snake;
},
```

3．随机食物位置

游戏中需要随机生成食物的位置，并且在刚创建蛇时，将蛇身每个点的位置与食物的位置进行比较，如果食物生成的位置在蛇身上，则重新生成，相关代码如下。

```
createFood() {
    //随机生成食物的位置
    //这里的20是背景高度(宽度)/格子高度(宽度),即 600 / 30 = 20
    this.food = {
        x: Math.floor((Math.random() * 20) + 1),
        y: Math.floor((Math.random() * 20) + 1)
    }
    for (var i = 0; i > this.snake.length; i++) {
        //创建蛇时,对蛇的位置与食物的位置进行比较
        var snakeX = this.snake[i].x;
        var snakeY = this.snake[i].y;
        //如果食物的位置出现在蛇身上,则重新生成
        if (this.food.x === snakeX && this.food.y === snakeY || this.food.y === snakeY && this.food.x === snakeX) {
            this.food.x = Math.floor((Math.random() * 20) + 1);
            this.food.y = Math.floor((Math.random() * 20) + 1);
        }
    }
},
```

4．碰壁

程序需要检测蛇头是否碰壁或者碰到蛇身，相关代码如下。

```
checkCollision(x, y, array) {
    for(var i = 0; i < array.length; i++) {
        if(array[i].x === x && array[i].y === y)
            return true;
    }
    return false;
},
if(snakeX == -1 || snakeX == this.w / this.snakeSize || snakeY == -1 || snakeY == this.h /
this.snakeSize || this.checkCollision(snakeX, snakeY, this.snake)) {
    //ctx.clearRect(0,0,this.w,this.h); //clean up the canvas
    clearInterval(this.interval);
    this.interval = null;
    this.restart()
    return;
}
```

5. 吃食物

程序需要判断是否吃到食物,吃到食物后将食物位置记录,分数加 5 并创建新的食物;如果没吃到食物,去掉数组最后的元素并返回,然后更新的蛇头位置,相关代码如下。

```
if(snakeX == this.food.x && snakeY == this.food.y) {
    //吃到食物
    //将食物的位置记录下来
    this.tail = {x: snakeX, y: snakeY};
    //分数加 5
    this.score = this.score + 5;
    //再创建食物
    this.createFood();
} else {
    //没吃到食物
    //去掉数组最后的元素并返回,相当于删除蛇尾
    this.tail = this.snake.pop();
    //将移动更新后蛇头的位置加到 tail 中
    this.tail.x = snakeX;
    this.tail.y = snakeY;
}
//unshift()方法可向数组的开头添加一个或多个元素
//将更新后的节点添加蛇头
this.snake.unshift(this.tail);
```

6. 完整逻辑代码

程序实现的完整逻辑代码请扫描二维码获取。

7.4 测试应用

本部分包括贪吃蛇小游戏 App 的程序调试和结果展示。

7.4.1 程序调试

项目编译成功后,DevEco 提供了几种调试方式可供选择。
(1) 真机调试。
(2) 远程真机调试。
(3) 模拟机调试。

本项目使用远程真机调试,其使用方法详见 HarmonyOS 官方文档:https://developer.harmonyos.com/cn/docs/documentation/doc-guides/ide_debug_device-0000001053822404。

7.4.2 结果展示

打开 App,应用初始界面如图 7-5 所示。

游戏开始时默认小蛇出生在左上方,单击屏幕任意地方即可进行游戏。通过屏幕下方的四个按键可以控制蛇头前进的方向,进行游戏,当小蛇吃到食物后,下方的积分系统会进行计分,一个食物是 5 分,如图 7-6 所示。

当小蛇碰到自己的身体或者碰触到四周墙壁时,游戏自动结束并暂停记分,如图 7-7 所示。

图 7-5 应用初始界面

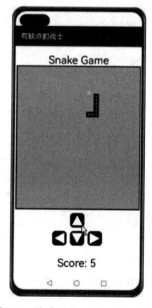
图 7-6 游戏运行界面及积分系统

图 7-7 游戏结束界面

7.5 问题解决

在程序开发过程中的主要问题有导入包报错、远程真机调试报错和 SDK 安装错误。

1. 导入包报错

问题描述：通过 gradle 导入包失败。

解决方法：通过外部 Module 形式导入，但由于 DevEco Studio 不支持直接导入外部 Module，需要将外部 Module 代码目录手动复制到自己的工程下，如图 7-8 所示。

图 7-8　导入外部 Module

修改工程下的 settings.gradle 文件，在里面添加该模块，修改后的配置为 include ':entry',':roundimage'。

在工程中，将要改变代码的 Module 导入模块依赖，修改 build.gradle 文件，添加依赖即可。

2. 远程真机调试报错

问题描述：使用远程真机调试时，无法自动配置签名，导致无法进行调试。

解决方法：进行手动签名，从 APPGallery Connect 中申请调试证书和 Profile 文件后，再进行签名。

3. SDK 安装错误

问题描述：创建项目后，提示 SDK 安装错误，错误显示为：Unable to download the HarmonyOS SDK. Unable to install toolchains：2.1.1.20，java：2.1.1.20 as the license has not been accepted。

解决方法：在菜单 Tools-SDK Manager 下安装所需要的 SDK 即可。

图 书 资 源 支 持

感谢您一直以来对清华大学出版社图书的支持和爱护。为了配合本书的使用，本书提供配套的资源，有需求的读者请扫描下方的"书圈"微信公众号二维码，在图书专区下载，也可以拨打电话或发送电子邮件咨询。

如果您在使用本书的过程中遇到了什么问题，或者有相关图书出版计划，也请您发邮件告诉我们，以便我们更好地为您服务。

我们的联系方式：

地　　址：北京市海淀区双清路学研大厦 A 座 714

邮　　编：100084

电　　话：010-83470236　010-83470237

资源下载：http://www.tup.com.cn

客服邮箱：tupjsj@vip.163.com

QQ：2301891038（请写明您的单位和姓名）

用微信扫一扫右边的二维码,即可关注清华大学出版社公众号。

教学资源·教学样书·新书信息

人工智能科学与技术
人工智能|电子通信|自动控制

资料下载·样书申请

书圈